ROUTLEDGE LIBR/
POLITICAL GE(

Volume 3

POLITICS, GEOGRAPHY & SOCIAL STRATIFICATION

POLITICS, GEOGRAPHY & SOCIAL STRATIFICATION

Edited by
KEITH HOGGART AND
ELEONORE KOFMAN

Routledge
Taylor & Francis Group

LONDON AND NEW YORK

First published in 1986

This edition first published in 2015
by Routledge
2 Park Square, Milton Park, Abingdon, Oxon, OX14 4RN

and by Routledge
711 Third Avenue, New York, NY 10017

Routledge is an imprint of the Taylor & Francis Group, an informa business

British Library Cataloguing in Publication Data
A catalogue record for this book is available from the British Library

ISBN: 978-1-138-80830-0 (Set)
eISBN: 978-1-315-74725-5 (Set)
ISBN: 978-1-138-80038-0 (Volume 3)
eISBN: 978-1-315-75548-9 (Volume 3)
Pb ISBN: 978-1-138-80039-7 (Volume 3)

Publisher's Note
The publisher has gone to great lengths to ensure the quality of this reprint but points out that some imperfections in the original copies may be apparent.

Disclaimer
The publisher has made every effort to trace copyright holders and would welcome correspondence from those they have been unable to trace.

MIX
Paper from
responsible sources
FSC
www.fsc.org   FSC® C013604   Printed and bound by CPI Group (UK) Ltd, Croydon, CR0 4YY

POLITICS, GEOGRAPHY & SOCIAL STRATIFICATION

Edited by
KEITH HOGGART and ELEONORE KOFMAN

CROOM HELM
London • Sydney • Wolfeboro, New Hampshire

Croom Helm Ltd, Provident House, Burrell Row,
Beckenham, Kent, BR3 1AT

Croom Helm Australia Pty Ltd, Suite 4, 6th Floor,
64-76 Kippax Street, Surry Hills, NSW 2010, Australia

British Library Cataloguing in Publication Data

Politics, geography and social stratification.
   1. Social classes 2. Anthropo-geography
   3. Social classes — Political aspects
   I. Hoggart, Keith II. Kofman, Eleonore
   305   HT609
   ISBN 0-7099-3784-9

Croom Helm US, 27 South Main Street,
Wolfeboro, New Hampshire 03894-2069

Library of Congress Cataloging-in-Publication Data

   Politics, Geography, and Social Stratification
   Includes index.
   1. Social classes — Great Britain — Congresses.
   2. Social structure — Great Britain — Congresses.
   3. Political sociology — Congresses. 4. Great Britain —
   Politics and government — 1945-   — Congresses.
I. Hoggart, Keith. II. Kofman, Eleonore.
HN400.S60P65 1986   305.5'0941   86-16539
ISBN 0-7099-3784-9

Printed and bound in Great Britain
by Billing & Sons Limited, Worcester.

CONTENTS

CONTENTS

TABLES AND FIGURES

Tables

TABLES AND FIGURES

Figures

ACKNOWLEDGEMENTS

The conference from which the papers in this volume are drawn was organised on behalf of the Social Geography Study Group of the Institute of British Geographers. We should like to thank the committee members of that study group for their advice and support in organising that conference. Also, much credit must be given to the Institute of British Geographers for providing financial assistance to help cover the expenses of the conference and for aiding poorly funded and non-funded graduate students to attend it.

The Scientific Section of the French Embassy, London, and in particular Agnes Pivot, provided advice and obtained financial support for us to enable Monique Pinçon-Charlot to present a paper at the conference.

In addition, the following publishers have kindly given their permission for material to be reproduced in this volume: Macmillan Press for Table 1 from Butler and Stokes <u>Political Change in Britain, The Evolution of Electoral Change</u>, 2nd edn., 1974; Cambridge University Press for Tables 2 and 3 from the article by Curtice and Steed in the <u>British Journal of Political Science</u>, 12, 1982, 129-98; Frances Pinter for Table 4 from McAllister and Rose, <u>The Nationwide Competition for Votes: The 1983 British Elections</u>, 1984; Oxford University Press for Table 7 from the book by McCallum and Readman <u>The British General Election of 1945</u>, 1947; and Methuen & Co. for Table 12 from Fogarty, <u>Prospects of the Industrial Areas of Great Britain</u>, 1947.

CHAPTER ONE

INTRODUCTION

KEITH HOGGART AND ELEONORE KOFMAN

The title of this volume reflects the dominant
themes within it and is indicative of the increasingly interdisciplinary thinking in the social
sciences. The material in this volume(1) was first
presented at a conference titled Geographical
Aspects of Social Stratification at which sociologists and geographers participated. Both the
papers and the discussions at the conference
repeatedly emphasised the political dimension and
implications of analyses of spatial differences
in social structuring. Thus the primary contribution of this volume to social research are the
links which are drawn between, first, the geographical basis of social stratification and social
relations and, second, the political outcomes of
socio-spatial differentiation. However, the set
of papers do not attempt to cover the entire gamut
of the subject matter of the various themes explored
in the volume. Instead, it seeks to present original papers which focus on particular aspects,
concepts and processes within socio-spatial structuring and its political consequences.
   To understand the origins of this collection
of papers, which span geography, politics and
sociology, it is necessary to highlight the growing
awareness of a need to relate sociologically based
research on social stratification with geographical
analyses of social differentiation. From the
sociological perspective, the theorising and investigation of social structures has predominantly
been national in orientation. This is largely
because the analysis of social classes and stratification has mainly been situated within macro-
sociology. It has been concerned with abstract
generalisations and the broad constraints of social
structures (for an attempt to bridge the micro

and macro-social theory gap, see Knorr-Cetina and
Cicourel 1981).

In a British context, this tendency has been
aided by the commonly ascribed representation of
British society as highly homogeneous in spatial
terms. Only recently have major regional social
divisions and regional political parties been
afforded much attention. The resurgence of
regionalist and nationalist movements, the more
sharply demarcated North/South political and economic cleavage, and the spatially variable performance of Britain's third (so-called 'centre')
political party, have all contrived to challenge
this image and emphasise the critical role of
location in socio-political differentiation. In
France a similar awakening of interest in the
geographical dimensions of social structuring has
occurred. Here, a research interest has developed
on the 'new' middle class (Bidou 1983; Benoit-
Guilbot 1985), which was a major element in bringing
the socialists to power in 1981, and particularly
on the political alliances they forged at the local
level. The importance of the geographical dimension
in these studies is brought out in the manner in
which the middle class defines itself through a
set of symbols, social practices and position in
the economic structure. Since these are inherently
heterogeneous, the same forces of unification or
production of (a middle) class does not hold in
all localities. The middle class as a socio-
political force has thereby been articulated in
quite distinctive ways across localities (see
Apkarian-Lacout and Verges 1983 for a study of
a communist municipality; and Verges 1983 for a
general analysis of class alliances in local
politics).

Thus both theoretical and political changes
have drawn attention to the need to incorporate
a geographical component into considerations of
social structures and social practices. As Giddens
has argued (1979, p.206):

> In class society, spatial division is a major
> feature of class differentiation. In a fairly
> crude, but nevertheless sociologically sig-
> nificant, sense, classes tend to be regionally
> concentrated...such spatial differentiation
> always have to be regarded as time-space
> formations in terms of social theory. Thus
> one of the important features of the spatial
> differentiation of class is the sedimentation

2

of divergent regional 'class cultures' <u>over time</u>.

Even given the growth in support for this line of argument, both in sociological (e.g. Urry 1981, 1983) and geographical contexts (e.g. Thrift 1983; Massey 1984), there is still a lack of concepts which make space, and the control over space, an integral part of social theory (Giddens 1981). There is a tendency amongst sociologists to belittle the relevance of spatial forms and not to realise that human beings 'make their own geography' as much as they 'make their own history' (Giddens 1984, p.363). This diminishes not only sociological research, but also social geography, which traditionally relied on sociology for its theoretical insights and, unfortunately, accepted its social categories uncritically (Jones and Eyles 1977). The comparatively late development of social geography as a subdiscipline, along with its leaning on sociology for its social concepts, has meant that sociology was best placed for linking social practices to geographical locations.

Further, within both sociological and anthropological research there is a long-established research tradition that explicitly focuses on locationally-specific social practices. This is the **community studies literature.** One might reasonably have expected that, either from Britain or the United States, this body of research would have provided some theoretical advancement in integrating geographical and sociological dimensions of social organisation and behaviour. However, there are two main reasons why this did not prove to be the case. First, there were two distinct, but coexisting, emphases in these studies. One group of analysts stressed the uniqueness of each locality. The other adhered to a belief in the universality of the type of society they portrayed in their particular community study (Tiévant 1983). These two emphases were never reconciled so that a comparative methodology, which might have lent itself to a dialogue with macrosociological approaches, was not forthcoming. Too often, then, the description of an individual locality was taken as a sufficient goal in itself. Theory seems to have been barely considered and, where it was, tended to focus on a straightforward comparison of national stratification structures with local ones (usually resulting in either a positive conclusion which stressed the strength of local social uniqueness or a negative one which decried the loss of community). At the same time there was a confusion over whether a community was being selected because it was an ideal observation unit or on account of its applicability to the wider social world.

The confusion that surrounded the aims of community studies was a reflection of a lack of clarity over the concept 'community' itself. In the main the concept of a community, as an analytical device or object of investigation, was not distinguished from the notion of community as an ideal living environment. Too often community was implicitly taken as a polar position in the conflict of two social orders. At one extreme was the industrial city, with its supposed alienated and utilitarian social relationships, while at the other end of the spectrum there was the traditional, rural community, with its camaraderie, homeliness and 'natural' habitat for people (Tiévant 1983, p.244). This dichotomy is clearly seen in the abstract in the normative overtones embodied in the concept of community (Bell and Newby 1971). More practically, it is evident in land use planners' desire to 'build' local residential communities and in over-sentimentalised accounts of both inner city neighbourhoods and isolated rural settlements (e.g. Day and Fitton 1975).

In effect, three major dimensions of the relationship between spatial and social structures have been weakly conceptualised; these are social class, space and time. Within the ambit of community studies, social class tended to receive comparatively scant attention. Overwhelmingly the community studies literature focused on social status divisions and tabulated, in an explicitly descriptive manner, the lifestyles of community residents (e.g. Willmott 1985). Even studies which could have been expected to adopt an overt class analysis tended to consider class relations more in passing than as a central theme (see, for example, investigations of coal mining communities like those of Dennis et al 1956 and Lantz 1958). So notable was this tendency that investigations which did highlight class relationships were readily distinguished by their content and conclusions (e.g. Littlejohn 1963). More commonly class was given scant attention or was primarily addressed to ascertain its interconnectedness with the social status and power dimensions of local social structures. Even the substantial literature that developed on community power structures rarely brought out obvious associations with local class relations. Instead this literature became bogged down in an acrimonious debate over the precise meaning of the concept 'power' and the most appropriate means of empirically investigating it (Walton 1976).

Whilst the work influenced by the Chicago School of Urban Ecology did concern itself with the relationship between spatial and social structures, this was postulated on an ecological form that assigned to each type of space or residential area a given cultural model. Thus, 'urban', 'suburban' and 'rural' represented distinctive ecological milieux. There was a tendency to contrast the homogeneous cultural models of the inner city and suburbia, although research in the 1960s did question this opposition by demonstrating the diversity of suburban life (Gans 1967). Castells argued, furthermore, that there was no inherent reason why a social unit should coincide with a spatial unit. This relationship ought to be investigated rather than assumed (Pickvance 1976, p.5; Castells 1977, p.108). In fact, such theorising of the spatial and the social has come to acquire importance in the weberian and marxist revival of urban sociology in the 1960s and 1970s.

As a further consideration, spatial and social structures are also interwoven in a time dependent matrix. The spatial manifestation of social practices reflect the historical legacy of restraints and opportunities which impinged on dominant social processes at the time of the late major transformation in a locality. Given that such local transformations occur at different times under different conditions, their social manifestations are dissimilar. Historically, we could compare the social upheavals of nineteenth century coal mining developments with those of North Sea oil. On the other hand, roughly similar patterns of de-industrialisation in mining areas of Wales and Cornwall have produced different effects because the process has been articulated with very different preexisting structures (Massey, cited by Urry 1983, p.38). These examples illustrate that the incorporation of a temporal dimension is essential if we are to understand the interplay of locality and nationwide social transformations, and how these determine differences in social practices and political conflicts within localities. In the main, however, the temporal dimension of social change - most especially when analysed to assess causal structures - was weakly integrated into the community studies literature.

The emergence of a more theoretised urban sociology pushed the relevance both of time and space in social processes more to the forefront. Both Rex and Moore (1967) and Pahl (1975), whilst not divorcing the city from its wider society, 'combined an emphasis on the sociological significance of spatial distributions of population and resources with the familiar weberian concerns with the goals and values of individual actors and the distribution of life chances in society' (Saunders 1981, p.118). The urban as an arena of struggle over consumption, and the emergence of social movements outside of the narrowly defined sphere of work, also demanded analysis from marxists. Yet their treatment of space and social processes reflected basic divergences between alternative marxist interpretations over the form of relationships between structures, human agency and political struggles. For **Castells**, for example, social processes simply unfold in space:

> A 'sociology of space' can only be an analysis of social practices given in a certain space... Of course there is the 'site', the 'geographical' conditions, but they concern analysis only as the support of a certain web of social relations (Castells 1977, p.442).

This interpretation stands in sharp contrast to those of Lefebvre and **Harvey**. For them, capitalist society has colonised, commodified and integrated space into its own social processes. Urban space provides a highly important means, for Lefebvre even a dominant means, of diverting and overcoming restrictions on the accumulation of surplus value (Lefebvre 1970, 1976; Harvey 1978). Thus the '... logic of capitalism is the logic of social use of space is the logic of everyday life. The class that controls production controls the production of space and hence the reproduction of social relations' (Saunders 1981, p.156). Space is above all a social product and not a natural entity. Thus conflicts over the use of space reveal contradictions of capitalism between private profit and social needs.

Whilst the new urban sociology of Castells highlighted the political potential of social movements, and indeed accorded prominence to those

6

engaged in struggles over collective consumption, it neglected broader non-specifically urban movements and those not directly involved in pursuing local class alliances. Its rigidly structuralist interpretation of urbanisation also provoked criticism for its implacable rejection of human agency and the concept of the individual subject (Saunders 1981; Whitt 1984). This left little room for understanding, say, the strategies which led to shifts in local alliances and the basis upon which the middle classes entered into alliances (the middle classes, in this context, being those to whom Wright (1979) ascribes a contradictory class position between the unambiguously defined bourgeoisie and the proletariat). Hence the social projects and strategies of the middle classes have been weakly theorised and barely researched as yet. Yet, the middle classes cannot be simply relegated to an untheorised terrain. Furthermore, the presence and impact of these middle classes are highly variable in different localities (Verges 1983; Silk 1984).

Castells' framework thereby stands in marked contrast to that of **Urry** (1981) who incorporates a geographical dimension into his account of the constitution and reproduction of human subjects in local civil societies. In demarcating civil society as being outside of both the sphere of production (economy) and the state, Urry acknowledges that all conflicts and bases of organisation cannot be reduced to capital-labour contradictions. This offers a more realistic framework in which to explore the complexity of relations and social practices. This is not to say that the appearance of non-class conflict does not have class relations as their fundamental underpinning (see Byrne, chapter 6). Within Urry's framework, however, a variety of dimensions to local social structure, and bases upon which individuals may mobilise, can be specified (p.70). In order of importance, these are:

1. Spatial organisation of labour and residence; that is, into nations, regions, cities, towns, countryside, neighbourhoods.
2. Sexual division of labour.
3. Religious/ethnic/racial allocation of subjects.
4. Differentiation of subjects on basis of trade-union and professional associations, artistic and leisure associations, political parties, neighbourhood groups, and various informal

groupings.
5.   Generational allocation of subjects.

According to Urry (chapter 2), the increasing
spatial and cultural fragmentation of class along
such dimensions is a key feature of a disorganised
capitalism which is currently evolving. This is
characterised by certain groups, such as women
and ethnic minorities, seeking to establish a fuller
equality that extends well beyond the formal
equality granted in the earlier period of capitalism
(Urry 1983, p.31). Viewed in this manner, social
mobilisation and collective action may not occur
along class lines. People might also be more
inclined toward informal participation rather than
pursuing goals through more formal political
channels (see Peake chapter 4).
    A debate is taking place, between those who
forsee breakdown in class structures and the emergence of new social forms, and those who conceptualise these occurrences as surface appearances
with class conflict still fuelling processes of
social structuring. Most overtly, this debate
has been played out in disagreements between those
adopting a weberian analytical posture and adherents
of marxism. Out of this debate has arisen
an increased awareness of the importance of new
dimensions of social structure and, whether from
a weberian or a marxist perspective, the need to
incorporate them into our theorising. How to do
so has been the subject of sharp, at times acrimonious, but certainly instructive dispute. In
this volume, the material incorporated reflects
some of the dimensions of this dispute. We have
not tried to impose a uniformity of view, nor a
uniformity of purpose on individual contributions.
From the outset our aim was to bring together
analysts who would focus on relationships between
spatial and social organisation from alternative
perspectives, utilising a variety of contexts to
illustrate their case. In a sense, therefore,
there is a certain eclecticism in this collection.
What binds it together is the significance of the
issues raised for investigations of socio-spatial
structuring and its political consequences.
    Incorporated with this book are papers whose
geographical focus ranges from advanced capitalist
societies as a whole to the home. There are empirical contributions and those that search for conceptual clarity. In places the geographical emphasis
is strong, in others the sociological dimension

is paramount. In all cases both are present, but the relative weighting authors have given them depends upon the particular 'story-line' they are weaving. This we feel is one of the strengths of their contribution, for the reader's ideas, questioning and understanding can feed off the varied approaches, empirical contexts and insights that each paper provides. To appreciate these differences, of course, some feel for the context in which these papers stand is required. This can be demonstrated by first looking at social class mobilisation, for this concept binds many of the contributors' insights.

It seems clear that not only is the mobilisation and identification of subjects along class and non-class lines a reality that depends on local systems of social stratification and interaction, but that it is also determined by the broader historical canvas that impinges on individual and class experiences. Class position and attendant social practices cannot be understood simply in terms of the present situation, for the past is present in the social trajectories of individuals, families and groups. Their experiences are re-fracted through what Bourdieu calls the habitus. By that he means the system of dispositions produced by the inherited experiences and structures of the social world which are internalised by the individual or group in the course of its existence (Bourdieu 1984, and see Pinçon-Charlot, chapter 8). The concept habitus enables us to make sense of how social hierarchies are reproduced and how they regulate social life even though individuals are not aware of these structures. Yet there is a problem with this concept, for within it the past weighs so heavily on individuals and groups that social change and innovation appears to be entrapped into solely reinforcing existing structures (Preteceille 1985, p.5.).

While we need to recognise that the past does impose severe constraints on behaviour, it is theoretically crucial to understand how major periods of social dislocation have engendered reappraisals of the social condition and political affiliations of individuals and social groups. Thus, for example, the Second World War expanded the horizons of millions of people, brought different social classes into greater daily contact and pushed many women into waged labour. One immediate consequence in Britain was the election of a Labour government (see Thrift, chapter

5). Many of the children of that generation experienced upward social mobility through the education
system. Their parents meanwhile were experiencing
rapid income increases, as the post-war economic
boom gained momentum in the 1950s. It was with
reference to this period that social scientists
began to seriously question the class basis of
British social organisation. This reflection formed
the basis of the so-called affluent worker thesis,
which was inspired by the apparent breakdown of
social class as a predictor of voters' party choices
in national elections (Goldthorpe and Lockwood
1968-9). To account for class dealignment in
national elections, it was suggested that a decline
had occurred in working class solidarity and social
mobilisation. This suggestion was based on the
premise that the link between social class and
party choice was uniform throughout the country.
As Alan Warde (chapter 3) shows, this is certainly
not the case. Local social structure is as important as national class position, as seen in the
manner in which the class composition of a constituency has increased in importance, and thereby
helped counteract the effect of dealignment at
the level of the individual voter. The full implications of this locality effect, and the mechanisms
of the production of a local political hegemony,
remain to be explored in detail.
    The origin of the decline in class identi-
fication at a local level has of course been
addressed by researchers like Dunleavy (1979) in
terms of consumption sector cleavages. Newby et
al (1985) present an alternative interpretation,
wherein class dealignment stems from the attitudes
and increasingly 'privatised' behaviour of the
affluent worker. This stands in contrast to the
communally-centred sociability of the (traditional
image of the) proletarian worker. The dichotomy
here is centred around the private sphere of the
home and the public arena of community and work.
As Newby et al warn, the conceptualisation of the
private should not be uniquely associated with
a retreat from the public world into the relatively
isolated household, for this process has developed
historically with the winning of citizenship rights
and integration of the individual into the nation
state. Yet, despite the acknowledged relationship
between the public and the private spheres, this
conceptualisation still overlooks the idea of the
home as a locale, or a context for social interaction (Giddens 1984, p.118). Even more significant

is the relationship between production and reproduction and the home as the site of social relations
and of socially defined roles and meanings (see
Williams chapter 11). The role of individual
members of the household and their relationship
to spheres of production and reproduction have
varied enormously both historically and between
social classes (MacKenzie and Rose 1983). Needless
to say studies of social stratification usually
assume that a household derives its class position
from the male head of a nuclear family. It is now
clear that the incorporation of women into wage
labour, along with the increase in home working,
needs to be considered in both social stratification
and conceptualisations of public/private realms
(see Peake, chapter 4). Thus, the struggle to
support local employment, as in Britain's year-
long miners' strike, and opposition to the privatisation of public services like health, have
begun to alter the collective and household practices of the women involved.

Of course, those excluded from work, male
and female, are not necessarily banished from the
public domain into an enclosed private domain.
What has happened, however, is that their contact
with the 'outside world' has increasingly come
to be mediated through state welfare agencies.
The rise of what Byrne (chapter 6) calls 'poverty
professionals' has been one feature of this process,
which has had clear implications for local class
restructuring. This direct involvement of state
agencies in restructuring the working class, and
in the creation of a marginalised section within
it, has placed state intervention more and more
into the regulation of social life. Marginalisation,
of course, is not simply the result of exclusion
from the labour market but also originates in the
operation of the housing market and in the inertia
it imposes on the inherited built environment (see
Hamnett and Randolph, chapter 9). Both of these
market processes concentrate the marginalised section of the population into the inner city. Yet
even in the inner-city there are few areas of total
class segregation.

Location provides certain common advantages
and disadvantages around which class alliances
can be forged. It is with reference to the integration of local alternative class structures and
locational attributes that Monique Pinçon-Charlot
(chapter 8) seeks to draw out associations between
social stratification, social practices and public

11

service provision.   The pertinence of this theme
to a British local government system is obvious.
Without understanding the mutual reinforcement
of class structure and the history of its restructuring in specific localities it is difficult to
appreciate the origins of local government
challenges to the present Conservative government's
policies (e.g. Duncan and Goodwin 1982).   Such
resistances are one aspect of a locationally
specific response to the restructuring of capital
at a national level. One dimension to this response
is the conscious striving to construct new relations
between the local state and labour and to support
oppressed groups within the labour force, particularly blacks and women (see Gough, chapter
7).   Such initiatives have recognised the heterogeneity of the working class in inner city areas
and the distinctive experiences and needs of such
groups.   Pressure for these types of initiatives
in and outside of the work place have emanated
from local political forces which sought to break
down, at least at the local level, the white male
ideology of the established political classes.
Yet, as Gough (chapter 7) makes clear, the existence
and promotion of such initiatives is highly constrained. Local deviations from dominant national
trends are difficult to maintain in times of economic hardship.   So, we should not ascribe a high
degree of autonomy to local authorities within
a system of capitalist hegemony, but acknowledge
that they can inflect national policies in distinctive ways (Preteceille 1981).   Nevertheless, the
decomposition of class structure arising from restructuring may result in different forms of local
social fragmentation within the working and middle
classes.   This can eventually lead to a process
of recomposition and new political alliances.
     In this volume, we have not sought to develop
a general framework which integrates all the diverse
elements of restructuring, class fragmentation,
social and political practices which converge around
areas of production/reproduction, public/private
spheres or formal/informal arenas. Nor have all
the papers addressed themselves directly to the
significance of space and localities in sustaining
social relations and their political impact. The
papers vary from general interpretations of economic
and social changes, which serve as a foundation
for empirical work on localities, through conceptual
and theoretical critiques which question the validity of current research practices, to empirical

evaluations of theoretical debates. As a conference collection, the volume is not intended as a rounded textbook, but as a vehicle for presenting thought-provoking insights on aspects of an increasingly important area of multidisciplinary social research.

REFERENCES

Apkarian-Lacout, B. and Verges, P. (1983) 'L'irrestible ascension des couches moyennes face à l'hégémonie communiste: Martigues', Sociologie du Travail, 25, 205-225.

Bell, C. and Newby, H. (1971) Community Studies, George Allen and Unwin, London.

Benoit-Guilbot, O. (1985) 'Avant propos', Sociologie du Travail (special issue on 'A propos des nouvelles couches moyennes'), 27, 119-121.

Bidou, C. et al (1983) Les couches moyennes salariées: mosaique sociologique, Ministère de l'Urbanisme et du Logement, Paris.

Bourdieu, P. (1984) Distinction: A Social Critique of a Judgement of Taste, Routledge and Kegan Paul, London.

Castells, M. (1977) The Urban Question: A Marxist Approach, Edward Arnold, London.

Day, G. and Fitton, M. (1975) 'Religion and social status in rural Wales: "Buchedd" and its lessons for concepts of stratifiction in community studies', Sociological Review, 23, 867-892.

Dennis, N., Henriques, F. and Slaughter, C. (1956) Coal is Our Life: An Analysis of a Yorkshire Mining Community, Eyre and Spottiswoode, London.

Duncan, S. and Goodwin, M. (1982) 'The local state: functionalism, autonomy and class relations, in Cockburn and Saunders', Political Geography Quarterly, 1, 77-96.

Dunleavy, P. (1979) 'The urban basis of political alignment: social class, domestic property ownership and state intervention in the consumption process', British Journal of Political Science, 9, 409-443.

Gans, H. (1967) The Levittowners, Allen Lane, London.

Giddens, A. (1979) Central Problems in Social Theory: Action, Structure and Contradiction in Social Analysis, Macmillan, London.

Giddens, A. (1981) A Contemporary Critique of Historical Materialism: Volume 1, Power, Property and the State, Macmillan, London.

Giddens, A. (1984) The Constitution of Society. Outline of a Theory of Structuration, Polity Press, Oxford.

Goldthorpe, J.H. and Lockwood, D. (1968-9), The Affluent Worker, 3 Volumes, Cambridge University Press, Cambridge.

Harvey, D. (1978) 'The urban process under capitalism: a framework for analysis', International Journal of Urban and Regional Research, 2, 101-31.

Jones, E. and Eyles, J. (1977) An Introduction to Social Geography, Oxford University Press, Oxford.

Knorr-Cetina, K. and Cicourel, A.V. (eds.) (1981) Advances in Social Theory and Methodology. Towards an Integration of Micro- and Macro-Sociologies. Routledge and Kegan Paul, Boston, Mass.

Lantz, H.R. (1958) People of Coal Town, Columbia University Press, New York.

Lefebvre, H. (1970) La révolution urbaine, Gallimard, Paris.

Lefebvre, H. (1976) 'Reflections on the politics of space' Antipode, 8, 30-37.

Littlejohn, J. (1963) Westrigg; The Sociology of a Cheviot Parish, Routledge and Kegan Paul, London.

MacKenzie, S. and Rose, D. (1983) 'Industrial change: the domestic economy and home life' in J. Anderson et al (eds.), Redundant Spaces in Cities and Regions, Academic Press, London, pp. 155-200.

Massey, D. (1984) Spatial Divisions of Labour, Macmillan, London.

Newby, H., Vogler, C., Rose, D. and Marshall, G. (1985) 'From class structure to class action: British working class politics in the 1980s' in B. Roberts et al (eds.), New Approaches to Economic Life: Economic Restrictions, Unemployment and Social Divisions of Labour, Manchester University Press, Manchester, pp. 86-102.

Preteceille, E. (1981) 'Left-wing local governments and services policy in France', International Journal of Urban and Regional Research, 5, 411-25.

Preteceille, E. (1985) 'Collective consumption, urban segregation and social classes' in 5th Urban Change and Conflict Conference, University of Sussex.

Rex, J. and Moore, R. (1967) Race, Community and

Conflict: A study of Sparbrook, Oxford University Press, Oxford.

Saunders, P. (1981) Social Theory and the Urban Question, Hutchinson, London.

Silk, J. (1984) 'Relationships between culture and society - a marxist proposal' I.B.G. Conference, University of Durham.

Smith, M.P. (1984) 'Urban structure, social theory and political power; in M.P. Smith (ed.), Cities in Transformation: Class, Capital and the State, Sage Urban Affairs Annual Review, 26, Beverly Hills, pp. 9-27.

Thrift, N.J. (1983) 'On the determination of social action in space and time', Environment and Planning D: Society and Space, 1, 23-57.

Tiévant, S. (1983) 'Les études de "communauté" et la ville: héritages et problèmes', Sociologie du Travail, 25, 226-232.

Urry, J. (1981) 'Localities, regions and social class', International Journal of Urban and Regional Research, 5, 455-473.

Urry, J. (1983) 'De-industrialisation, classes and politics' in R. King (ed.), Capital and Politics, Routledge and Kegan Paul, London, pp. 28-48.

Verges, P. (1983) 'Approche des classes sociales dans l'analyse localisée', Sociologie du Travail, 25, 243-256.

Walton, J. (1976) 'Community power and the retreat from politics: full circle after 20 years?', Social Problems, 23, 292-303.

Whitt, J.A. (1984) 'Structural fetishism in the new urban theory', in M.P. Smith (ed.), Cities in Transformation: Class, Capital and the State, Sage Urban Affairs Annual Review 26, Beverly Hills, pp. 75-89.

Willmott, P. (1985) 'Making sense of everyday lives', Times Higher Education Supplement, 15 March, 16.

Wright, E.O. (1979) Class, Crisis and the State, New Left Books, London.

NOTES

1. All the papers, with the exception of Newby et al (1985) are included in this volume. The latter has already been published in Roberts et al (1985) New Approaches to Economic Life.

CHAPTER TWO

CLASS, SPACE AND DISORGANISED CAPITALISM(1)

JOHN URRY

Social stratification can be understood in two
ways, either in terms of the distribution of various
kinds of resources to individuals or groups, or
in terms of the social relations between groups
from which the distribution of resources can be
seen to follow. Sometimes, however, these two
forms of analysis are not sufficiently distinguished
and it is held that the unequal distribution of
resources provides the explanation of how and why
certain forms of social relations occur between
differentially stratified social groupings.
      In the following I shall assume that the un-
equal distribution of resources does not explain
the social relations occurring between social
groups, and that those relations are accountable
for in terms of complex processes involving the
constitution and reproduction of places in the
social division of labour. I shall also assume
that the processes by which such places are constructed and reproduced partly depends upon the
persons recruited to fill such places and the degree
to which they are able to organise to close off
entry into or encourage exit from sets of such
places. However, my main point here is that given
that a particular set of persons occupy a given
set of places, then whether those persons engage
in collective action or not with regard to the
distribution of resources, is highly problematic.
It is by no means the case that simply because
a particular group shares some condition in common
then those persons will organise to do anything
about that common condition. I shall now briefly
elaborate some of the recent debates about the
determinants of such collective action. Two points,
though, are worth bearing in mind since I shall
return to them: first, the spatial organisation

of social groupings plays a significant role in accounting for those different patterns of collective action; and second, it is necessary to consider in detail the periodisation of capitalist societies and especially how the contemporary growth of 'disorganised' capitalism transforms both the causes and consequences of collective action.

The starting point for all discussions of this issue is the so-called prisoner's dilemma game. The paradox of this game is that if both prisoners pursue their rational self-interest they end up with a result that is less satisfactory than if they had in some way been able to sacrifice those individual interests (for details of the game and some of the critical issues see Barry and Hardin (1982). This contradiction between what Barry and Hardin (1982) call 'rational man' (sic) and 'irrational society' involves a radical critique of the argument that self-interest is necessarily rational overall. The prisoner's dilemma game demonstrates that such self-interest may systematically fail to realise the social good. This argument has been used by Olson (1965) to show that there are important parallels between the situation of a firm in a competitive market and that which faces many individuals and groups confronted by choices and decisions about whether to join or participate in various sorts of organisation (see also Barry 1970; Heath 1976; Barry and Hardin 1982). Olson says that:

> A lobbying organisation, or indeed a labor union or any other organisation, working in the interest of a large group of firms or workers in some industry, would get no assistance from the rational, self-interested individuals in that industry (p.11).

The reason for this derives from the problem of the so-called 'free-rider'. Where the group to be organised is large and where the benefits from such organisation are public and cannot be confined to particular individuals, then the group is latent and will fail to be so organised unless individuals are induced to cooperate through the provision of other non-collective (selective) incentives. Without such selective benefits, individuals can 'free-ride', gaining the general benefits of the organisation if any materialise, but not incurring any of the material, temporal or motivational costs of membership.

17

Olson maintains that large groups are likely to remain 'latent' since they have particular problems in establishing and sustaining collective organisation. They will be less able to prevent free-riders, particularly because there will be little development of the social pressures and beliefs that would otherwise induce commitment to the organisation in question. In a large group not everyone can possibly know everybody else and so each person will not ordinarily be affected if he or she fails to make sacrifices on the group's behalf. Furthermore, the advantages to the whole that each person's contribution can bring cannot but be fairly slight - hence it will not seem worthwhile for a given individual to contribute time, energy or money to the organisation in question. At the same time, the larger the organisation, the greater the costs involved in getting it off the ground and into a form whereby any of the collective goods can be obtained. This applies particularly to the problems of mobilising large-scale social forces, such as classes.

There are a number of difficulties in Olson's argument. It has been pointed out that 'selective benefits' cannot explain the enormous diversity and strength of actually occurring collective organisations. This is especially relevant to explaining the existence of many groups which do not directly produce any material gains for their members, what Heath (1976) terms 'altruistic' pressure groups. These demand analysis of both 'moral incentives' and 'political entrepreneurship', which can both partly neutralise the otherwise pervasive 'free-rider' effect. However, the most serious problem in Olson's analysis concerns the assumption of independent individual self-interest that is involved. Barry points out that social life is analogous, not so much to a single-play prisoner's dilemma, but rather to an iterated prisoner's dilemma or 'supergame' (Barry and Hardin 1982, pp. 19-37). Where there is iteration the two players who should rationally defect (not co-operate), may in fact cooperate. This is because people often play over time what are in effect a great number of prisoner's dilemma games. They may gradually perceive that if they pursue their narrow self-interest they end up with a non-optimal collective solution. Hence, there is a learning process by which many of the actors try out solutions to the games they play which are individually non-rational. If these actors then come to realise there are major collective gains which result from the pursuit of individually non-rational solutions, then these cooperative games may become institutionalised amongst many of the players involved. It thus becomes rational to engage in social practices which would ensure such agreement. Indeed, to the extent that some agents operate non-rationally then there are no obvious predictions possible from Olson's argument. This is because the construction of rational choice models of this sort requires not merely that all agents within a given category act rationally but that they act rationally in essentially the same way. If they do not, then the predictions from the theory are indeterminate.

The most recent and illuminating analysis of such a theory can be found in Jon Elster's writings (Elster 1978, 1982; also Lash and Urry 1984). He argues inter alia for the importance of two basic premises of rational choice theory: (1) that structural constraints do not completely determine the actions individuals take; and (2) that within the feasible set of actions compatible with such constraints, and possessed with a given 'preference structure', an individual will choose those that he or she believes will bring the best results. Indeed, humans are so rational that even in situations where they are only imperfectly rational (e.g. in smoking, eating, etc.), they are able to bind themselves to future courses of action which will enable them to realise these desired goals (to stop smoking, lose weight, etc.). Analysis of such rational choices involves game theory, particularly because of the necessity to investigate the interdependence of decisions. Elster (1982, p. 464) argues that game theory has a particular contribution to make to marxist social science, 'because classes crystallise into collective actors that confront each other over the distribution of income and power, as well as over the nature of property relations; and as there are also strategic relations between the members of a given class, game theory is needed to explain these complex interdependencies'. In particular, in Logic and Society (1978), following his analysis of various kinds of social contradiction, Elster maintains that classes are more likely to develop collective action: (1) the more that actors perceive that there is some kind of contradiction characterising the society within which they are implicated; (2) the lower the 'communicational distance'

19

between the members; (3) the less the rate of turnover in group membership; and (4) the greater the degree to which contradictions are reversible (Elster 1978, pp. 134-150).

Charles Taylor (1980) has suggested in criticism that this kind of analysis is only relevant to societies with 'atomistic forms of life', since Elster is unaware of the necessity of community, of the norms of a Sittlichkeit, the 'common meanings' which make society possible. A further problem is that Elster defines class consciousness operationally as the capacity of a class to overcome the 'free-rider' problem. However, this would translate empirically into the incorrect proposition that the Swedish and Austrian working classes are the most class conscious in the West, and the French the least class conscious. What Elster ignores are the diverse ideological conditions that cannot be simply reduced to whether the members of different classes are or are not in close interaction with one another. In the USA, for example, the widespread existence of individualistic ideologies appears to lower union membership and to maximise prisoner's dilemma preference structures by comparison with the UK or Italy. Finally, although Elster elaborates the importance of the interdependence between classes (and presumably other social forces) he views them as essentially comprised of individuals who may or may not engage in collective action. What is missing from his analysis is an examination of classes as comprising sets of 'resources, capacities, and powers' which may be realised within specific conjunctures. It is these resources, capacities and powers which are crucially relevant to the consideration of whether a particular class can be collectively organised, and to the variable consequences of such organisation both on the class in question and upon other social forces within that society (Lash and Urry 1984).

This analysis clearly depends upon some prior conception of the social structure, of the relationships between social entities. This aspect is particularly developed within the related arguments of Offe and Wiesenthal (1980), who investigate the different organisational forms of labour and of capital. They argue that the crucial feature of labour is its individuality, it is atomised and divided by competition; moreover, labourers cannot merge, merely associate. Also, because of the indissoluble links between labourers and their labour-power, associations of labour must organise a wide spectrum of the needs of labour. Capital, by contrast, is united and is merely organised to maximise profits, this being a matter which can generally be left to decisions by technical experts. And at the same time, labour has to concern itself far more systematically with the well-being of capital, than does capital have to concern itself with the conditions of labour. Offe and Wiesenthal thus demonstrate that the associations of labour are defensive, they are responses <u>to</u> the collective organisation of capital. The latter may organise further in response to the associations of labour, either in informal cooperation or employers' associations, sometimes mediated via the state. Thus capital possesses three forms of organisation, the firm itself, informal cooperation between firms, and the employers' association - labour by contrast has merely one. In any conflict between the two capital would seem certain to win, since their collective action involves far fewer individuals, they are more united, and they possess clearer goals and greater resources.

Offe and Wiesenthal thus maintain that for the associations of labour to be viable an alternative organisational form has to develop - what they term the 'dialogical'. This involves, not merely aggregating the individual resources of the association members to meet the common interests of that membership, but also and more distinctively, defining a <u>collective identity</u>. Labour can only transform existing relationships by overcoming the relatively greater costs of engaging in collective action, as compared with capital. And this can only be achieved by deflating the standards by which such costs are assessed within their collectivity. The establishment of this collective identity is essential since it is the only means by which the subjective deflation of the costs of organisation can be effected. Moreover, it is only labour who may develop this non-utilitarian form of collective action, a form in which it is held that the costs of membership of the organisation are not to be assessed instrumentally. Also the organisations of labour rest upon the 'willingness to act', those of capital on the 'willingness to pay'. For the latter, then, there is no problem involved in maximising size - for the former this generates profound dilemmas. This is partly because an increase in size will increase the heterogeneity of members' occupations and interests, and hence will make it more difficult to establish the collective identity necessary for common action. The larger the organisation the more heterogeneous are the interests that have to be reconciled - not merely those of maximising members' wages, but also of ensuring security of employment, some control over the work process, and of pleasant working and living conditions. Unlike organisations of capital, which can create and maintain the integration of their membership in a one-dimensional 'monological' manner, organisations of labour are involved in a complex and contradictory process of expressing/forming/sustaining a common identity - an identity which cannot be assessed in purely instrumental terms. The power of capital exists without organisation, the power of labour only exists with organisation, but it is an organisation which is precariously established. The organisation in part has to function 'dialogically', whereby the activity and views of the membership have to be represented and embodied so as to sustain the necessary collective identity. Thus Offe and Wiesenthal argue that the organisations of capital are 'monological', those of labour have to be both 'monological' and 'dialogical'. They also argue that, compared with labour, the interests of capital are less ambiguous, controversial or likely to be misperceived. They do not require dialogical organisations in order to identify and to sustain such interests. Offe and Wiesenthal suggest that in recent decades the monological has increasingly come to replace the dialogical as the predominant organisational form of labour. The 'free rider' issue is secondary; after a certain organisational size is reached, it is not increased membership, but the consequential creation of dialogical forms, which is the main problem for sustaining collective action.

One limitation of Offe and Wiesenthal's otherwise insightful analysis is that they do not consider the external conditions under which labour may or may not develop a dialogical form of organisation. In particular, they do not consider the spatial structuring which may facilitate the establishment of the sustaining of the 'dialogical' and hence of collective action. I shall now set out a revised version of Elster's conditions for collective action which does take this seriously into account (see also Cannadine 1983; Lash and Urry 1984). A working class is more likely to engage in collective action, the more that:

(i)   there are a number of spatially specific
       but overlapping and class-based col-
       lectivities-in-struggle' in which there are
       shared and reasonably long-established experi-
       ences at work and/or through residential
       propinquity which facilitate the establishment
       of 'dialogue'.

(ii)   the spatially separated experiences of dif-
       ferent groups of workers can each be viewed
       as representing the experiences of the whole
       class within a given nation-state.   This
       depends upon a wide number of local 'civil
       societies' being structured by the class
       division between capital and labour rather
       than by more complex class/status divisions
       or by the division between the people versus
       the state.   Minimally in each locality this
       implies that (a) there is a degree of resi-
       dential differentiation between classes and
       (b) those residentially differentiated classes
       are nevertheless spatially adjacent.

(iii) other collectivities within local 'civil
       societies' are organised in ways which either
       reinforce these class divisions or are at
       least neutral with respect to them.   Col-
       lective action is thus more likely the more
       that civil society is organised on a 'ver-
       tical' basis in which there are few social
       groupings and other social practices which
       are non-class specific.

(iv)   gains and benefits (such as higher incomes,
       lower prices, increased educational and other
       opportunities, improved housing, better con-
       ditions or work, and so on) are thought to
       be and are as a matter of fact unavailable
       except through collective action of a broadly
       class-based sort.   This condition will be
       more likely to be met where social inequali-
       ties result from a nationally-based system
       of class relations, in which in a clear sense
       class relations produce the major dimensions
       of social inequality.

(v)   a substantial proportion of workers within
       a variety of different spatial locations
       conclude that class actions can be successful
       and are therefore worth pursuing, even
       if they do not immediately produce successful
       results; also, that residential and work
       patterns are such that levels of participation

ensure that success does not come to be measured simply in monological/qualitative terms.

Thus far my analysis has been abstracted from specific struggles in specific societies. I now want to begin to think out how to make the analysis less abstract by considering how to periodise capitalist societies. This is necessary because the specific causes and consequences of collective action vary between different societies, and that a first approach to the analysis of such variety involves the attempt to specify different forms of capitalist society, that is, the clear and identifiable interrelationships between most of a society's constitutive elements (economy/politics/civil society/the state and so on).

Existing periodisations in the social science literature are peculiarly unsatisfactory. In the non-marxist literature the main distinction developed has been that between pre-industrial and industrial society. It is then argued that there is a process of convergence towards the single industrial society particularly because of the mobilising power of advanced technology (see Kerr et al 1962). The main critiques of this have either been to emphasise the diversity of industrial capitalist societies and the resulting patterns of divergence; or to maintain that contemporary capitalism at least in some countries has become 'de-industrialised' or post-industrial (Touraine 1974). In the marxist literature the main distinction drawn has been between competitive and monopoly capitalism, with a further stage of state monopoly capitalism or late capitalism added later (Mandel 1972; Poulantzas 1975).

These formulations are unsatisfactory since they each suffer from the following deficiencies: (1) they are economistic in that the crucial property is held to be the economy (either in the form of technology or in terms of monopolistic, strictly oligopolistic, control of markets etc.), and other social institutions and practices are presumed to take their characteristics from it; (2) they do not take sufficient account of the fact that capitalist social relations are necessarily embedded within nation-states so that what may be true at the level of the world economy (increasing monopolisation, for example) is not necessarily an accurate reflection of developments within a localised nation-state (where there may be increased competition as monopolies relocate abroad); (3) they do not take into account the nature of class and other struggles occurring within such nation-states, with whether struggles are localised or 'nationalised', with the 'capacities' and 'resources' of classes and other social groupings, with the forms of state mediation, and so on.

I shall now briefly summarise the differences between what I shall term 'organised' and 'disorganised' capitalism, in each case in a manner which hopefully does not repeat the errors just elaborated. The notion of organised capitalism has a considerable pedigree dating back to Hilferding and was particularly developed by Jurgen Kocka, Heinrich Winkler, Hans-Ulrich Wehler and other social historians at a conference in 1972 (see Winkler 1973). For these writers organised capitalism begins in most countries in the final decades of the nineteenth century as a consequence of the downward phase of the Kondratieff long wave which began in the mid-1870s. According to these writers, organised capitalism consists of the following interrelated features:

1. concentration and centralisation of industrial, banking and commercial capital - as markets became progressively regulated.
2. growth of the separation of ownership from control, with the bureaucratisation of the latter and the elaboration of complex managerial hierarchies.
3. growth of new sectors of managerial/scientific/technological intelligentsia and of a bureaucratically employed middle class.
4. the growth of collective organisations in the labour market, particularly of regionally and then nationally organised trade unions and of employers' associations, nationally organised professions etc.
5. increasing inter-articulation between the state and the large monopolies; and between collective organisations and the state as the latter increasingly intervenes in social conflicts.
6. imperialist expansion and the control of markets and production overseas.
7. changes in politics and the state: including the increasing number and size of state bureaucracies, the incorporation of various categories into the national political arena,

the increased representation of diverse interests in and through the state, and the transformation of administration from merely 'keeping order' to the attainment of various goals and national objectives.

8. ideological changes concerning the role of technical rationality, the glorification of science and the significance of the 'national interest'.

The following are also to be seen as aspects of organised capitalism and need to be added to the formulation just presented:

9. the concentration of industrial capitalist relations within a relatively few industrial sectors and within a small number of centrally significant nation-states.

10. extractive/manufacturing industry as the dominant sectors with a relatively large number of workers employed.

11. a concentration of different industries within different regions, so that there are clearly identifiable regional economies based on a handful of centrally significant extractive/manufacturing industries.

12. increasing numbers employed in most plants as the economies of scale dictate growth and expansion within each unit of production.

13. the growth and increased importance of massive industrial cities which dominate particular regions through the provision of centralised services (especially commercial and financial).

Clearly not all of these developments occurred either simultaneously or in the same way in all countries in western Europe. It is useful to distinguish between organisation 'at the top' and organisation 'at the bottom'. German capitalism was organised early on at both the top and the bottom (1873-95), American capitalism was organised fairly early on at the top but very late on and only briefly at the bottom, Swedish capitalism was organised in the inter-war period at both the top and the bottom, French capitalism was only fully organised at top and bottom during and after the Second World War, while Britain was organised only very late at the top but rather early on at the bottom.

The following are three factors which determine the timing of, and the extent to which, the capitalism of a given country becomes organised. First, is the point in history at which it begins to industrialise. The earlier a country enters into its 'take-off', the less organised mutatis mutandis its capitalism will be. This is because countries which are later industrialisers need to begin at higher levels of concentration and centralisation of capital to compete with those which have already been industrialising for some time. Secondly, there is the extent to which pre-capitalist organisations survive into the capitalist period. On this count, Britain and Germany became more highly organised capitalist societies than France and America because the former two nations did not experience a 'bourgeois revolution'. As a result, guilds, highly corporate local government, and merchant, professional, aristocratic, university and church bodies remained relatively intact. Sweden interestingly occupies a mid-way position, in as much as the high level of state centralisation during Swedish feudalism did not allow for the same flourishing development of corporate groups. The third factor is the size of the country. For the industry of small countries to compete inter-nationally, resources had to be channelled into relatively few firms and sectors. Co-ordination between the state and industry was then greatly facilitated if not necessitated. At the same time there would tend to be higher union densities where there were relatively few firms and sectors (as in Sweden).

I will now consider briefly what is meant by 'disorganised capitalism', which I would claim characterises Britain since the 1960s and will characterise most other western societies, Sweden and Germany being the slowest to disorganise. The following are the features of disorganised capi-talism (the numbers correspond to those above):

1. the growth of a world market combined with the continued concentration and centralisation of capital means that national markets have become less regulated by nationally-based corporations. This has been heightened by the general decline of tariffs and the en-couragement by states to increase the size and external activities of large corporations. Increased centralisation now means increased competition.
2. the continued expansion of the number of white collar workers and particularly of a

distinctive service class (of managers, professionals, educators, scientists etc.) which
is an effect of organised capitalism becomes
an increasingly significant element which
then disorganises modern capitalism. This
results both from the development of an edu-
cationally-based stratification system which
fosters individual achievement and mobility
and the growth of new 'social movements'
(students, ecological, women's movements,
etc.) which increasingly displace class politics.

3. decline in the absolute and relative size
of the core working class, that is, of manual
workers in manufacturing industry, as economies
are de-industrialised.

4. decline in the importance and effectiveness
of national bargaining procedures and the
growth of local arrangements especially via
bargaining by shop stewards.

5. increasing independence of large monopolies
from direct control and regulation by individual nation-states; the breakdown of most
neo-corporatist forms of state regulation
of wage bargaining, planning, etc., and increasing contradiction between the state and
capital (cf fiscal crisis, etc.).

6. the spread of capitalism into most Third World
countries which has involved increased competition in many of the basic extractive/
manufacturing industries (steel, coal, oil,
heavy industry, automobiles etc.) and the
export of part of the first world proletariat.
This in turn has shifted the industrial/occu-
pational structure of first world economies
towards 'service' industry and occupations.

7. increasing alienation from national-based
parties which in some cases at least represent
class interests. There is a very significant
decline in the class vote and the more general
increase in 'catch-all' parties which reflect
the decline in the degree to which national
parties simply represent class interests.

8. an increase in cultural fragmentation and
pluralism, resulting both from the commodification of leisure and the development of
new political/cultural forms since the 1960s.
The growth of 'figural' rather than simply
'textural' bases of regulation has permitted
a decodification of existing cultural forms.
The related reductions in time-space

distanciation (cf. the 'global village') like-
wise undermine the construction of unprob-
lematically national class subjects.

9. the massive expansion in the number of nation-
    states implicated in capitalist production
    and the large expansion in the number of sec-
    tors organised on the basis of capitalist
    relations of production.

10. decline in the absolute and relative numbers
    employed in extractive/manufacturing industry
    and in the significance of those sectors for
    the organisation of modern capitalist
    societies. Increased importance of service
    industry for the structuring of social re-
    lations (smaller plants, less changeable labour
    process, increased feminisation, higher 'men-
    tal' component, etc.).

11. the overlapping effect of new forms of the
    spatial division of labour has weakened the
    degree to which industries are concentrated
    within different regions. To a significant
    extent there are no longer 'regional economies'
    in which social and political relations are
    formed or shaped by a handful of centrally
    significant extractive/manufacturing indus-
    tries.

12. decline in average plant size because of shifts
    in industrial structure, substantial labour-
    saving capital investment, the hiving off
    of various sub-contracted activities, the
    export of labour-intensive activities to
    'world-market factories' in the Third World,
    and to 'rural' sites in the first world, etc.

13. industrial cities begin to decline in size
    and in their domination of regions. This
    is reflected in the industrial and population
    collapse of so-called 'inner cities', the
    increase in population of smaller towns and
    more generally of semi-rural areas, the move-
    ment away from older industrial areas, etc..
    Cities also become less centrally implicated
    in the circuits of capital and become progress-
    ively reduced to the status of alternative
    pools of labour-power.

I have so far argued that (1) social inequalities do not in themselves generate collective
action, (2) collective action only results when
a number of spatial (and other) conditions hold,
(3) recent developments with the first world have
resulted in a new form of 'disorganised' capitalism.

29

In conclusion, I shall briefly consider the significance of such 'disorganisation' for the collective action of the working class returning to the five conditions outlined earlier.

The contemporary working class has enormous difficulties sustaining collective action because:

(i)   the radical restructuring of modern industry and policies of residential relocation have undermined some of the conditions facilitating sustained 'dialogue', especially <u>across</u> localities (the case of the miners surely being the exception that 'proves' the rule).

(ii)   it is now less true that local 'civil societies' are simply structured by divisions of social class - the division between the 'people' and the 'state' being an increasingly significant underlying social relation. Moreover, the growth of 'low cost' ownership of especially new homes has broken down some patterns of residential differentiation between social classes; while at the same time the internationalisation of capital and the progressive separation of ownership and control has reduced the degree to which there are locally resident 'bourgeoisies' which are both spatially differentiated and yet spatially adjacent.

(iii) since the 1960s there has been a massive expansion in the number and range of voluntary associations and socio-political groupings (the 'fragments'); and at the same time of the significance and impact of the means of mass entertainment. Both developments result in the 'horizontal' disarticulation of civil society.

(iv)   class relations do not appear to cause the entire patterning of social inequalities, particularly because of the 'apparently' increased importance of the division between the state and the people, gender divisions, age differences, and relations between ethnic groups. All of these sources of social inequality generate forms of non-class collective action which can provide bases for achieving gains and benefits.

(v)   clearly very many social groupings have concluded that collective action can be successful and is therefore worth pursuing. This stems from the increased horizontal fragmentation of civil society, the so-called 'breakdown of deference', and the development of a cultural pluralism, particularly focused around the 'personal', the 'body' and the 'natural'. But at the same time, for many social groupings which are work-based, it is hard indeed to ensure that 'dialogue' is sustained and that retreat into the monological or sheer opportunism does not gather pace.

To the extent, therefore, that western societies either are or are becoming to varying degrees 'disorganised', then the patterns of struggle and conflict are being redrawn partly in relationship to the changing spatial parameters. Such societies are hence transformed as new patterns of struggle and conflict work their way through the society and through the distribution of individuals and groups in time-space. The relations between class and space are being fundamentally restructured as the advanced capitalist societies are becoming disorganised.

NOTES

1. Many of the arguments here have resulted from collaborative work with Scott Lash. This chapter represents part of our 'work in progress'. We are currently preparing The End of Organised Capitalism Polity, Cambridge, 1986.

REFERENCES

Barry, B. (1970) Sociologists, Economists and Democracy, University of Chicago Press, Chicago.

Barry, B. and Hardin, R. (eds.) (1982) Rational Man and Irrational Society, Sage, Beverly Hills.

Cannadine, D. (1982) 'Residential segregation in nineteenth century towns: from shapes on the ground to shapes in society' in J. Johnson and C. Pooley (eds.), The Structure of Nineteenth Century Towns, Croom Helm, London, pp. 235-251.

Elster, J. (1978) Logic and Society, John Wiley and Sons, Chichester.

Elster, J. (1982) 'Marxism, functionalism and game theory: the case for methodological individualism', Theory and Society, 11, pp. 435-482.

Heath, A. (1976) Rational Choice and Social Exchange, Cambridge University Press, Cambridge.

Kerr, C., Dunlop, J.T., Harbinson, F.H. and Myers, C.A. (1962) Industrialisation and Industrial Man, Heinemann, London.

Lash, S. and Urry, J. (1984) 'The new marxism of collective action: a critical analysis', Sociology, 18, 33-50.

Mandel, E. (1972) Late Capitalism, New Left Books, London.

Offe, C. and Weisenthal, H. (1980) 'Two logics of collective action: theoretical notes on social class and organisational form' in M. Zeitlin, (ed.), Political Power and Social Theory, JAI Press, Greenwich, Connecticut, pp. 67-115.

Olson, M. (1965) The Logic of Collective Action, Harvard University Press, Cambridge, Massachusetts.

Poulantzas, N. (1975) Changes in Contemporary Capitalism, New Left Books, London.

Taylor, C. (1980) 'Formal theory in social science', Inquiry, 23, pp. 139-144.

Touraine, A. (1974) The Post-Industrial Society, Wildwood House, London.

Winkler, H. (ed.) (1973) Organisierter Kapitalismus, Vandenhoeck and Ruprecht, Gottingen.

CHAPTER THREE

SPACE, CLASS AND VOTING IN BRITAIN

ALAN WARDE

Introduction

There has always been a considerable degree of
spatial variation in voting patterns in Britain.
This is demonstrable in many ways. No one could
deny that Northern Ireland and Wales have per-
sistently deviated from the average patterns of
the UK throughout the 20th century. Scotland,
of late, has started to produce quite distinctive
voting behaviour. Even within England some regions
have shown a persistent preference for particular
parties which do not conform to the national average. A simple expression of regional variation
is the distribution of seats won by each party
during the 20th century (see Figure 1).

For a number of reasons this variation has
not been considered of much interest. In international perspective, Britain's regional variations
are relatively subdued. Britain has been considered
one of the most homogeneous of political societies
in the Western world (Rose and Urwin 1975). Spatial
homogeneity appeared to be proven by a number of
empirical observations. Until the revival of
nationalisms in the Celtic periphery in the 1960s
there were no regional parties. Within England,
regional differentiation in social and industrial
structure was thought minimal. No strong regional
identifications were apparent. The uniform national
swing at general elections seemed to entail that
voting was responsive to the politics of the centre
rather than of locality or region. Most of all
it was asserted that British electoral behaviour
was almost entirely a function of social class.
Britain was the archetypal 'modern' polity in the
sense that functional cleavages had replaced all
others. Partisanship was an effect of the

Figure 1: Regional distribution of parliamentary
seats 1922-1983 election

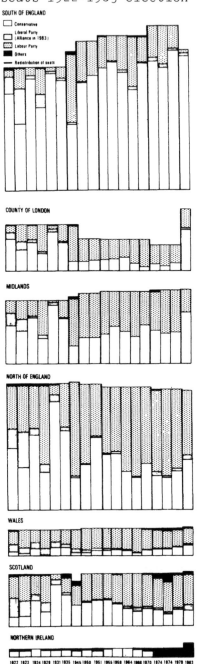

Source: D. Butler and A. Sloman (1980)

institutionalisation of class conflict. In such circumstances, spatial variation in voting behaviour was attributed directly to the spatial distribution of social classes. That is to say, the different results across constituencies were a direct effect of industrial and occuaptional structure. Interestingly, these last two reasons are, prima facie, inconsistent. If voting is a direct function of an individual's social class, and social classes are unevenly distributed geographically, then the swing between elections should not be uniform. This, the 'paradox of uniform swing', provided the principal focus of interest in spatial effects in Britain.

Interest in the geography of elections has to some degree revived recently. The increasing success of parties other than Labour and Conservative has led to an interest in the spatial variations in their support. Recent elections have also shown the existence of spatial trends in the support for the two major parties, the most obvious instance being the geographical concentration of Labour support in the 1983 election. Furthermore, an individual's social class is no longer a very good predictor of his or her vote, as is shown by survey analysis over the last decade (see below). There is a paradox surrounding the 'class dealignment thesis', though: while class has declined as a predictor of an individual's vote, the class composition of a constituency is a remarkably effective predictor of the election result in a constituency. This paradox, 'Miller's paradox' (Miller 1978), is the main subject of this paper.

Finally, by way of preface, it should be pointed out that political events, policy preferences and party identification affect how people vote! Any presumption that there will be a constant correlation between social or economic conditions and voting behaviour is absurd. Nor should voting be fetishised. Sociologists have paid limited attention to voting in Britain because, quite correctly, they reasoned that the most important issues of power in British society were unlikely to be tapped through examining electoral behaviour. Nevertheless, voting does offer an interesting challenge to sociological analysis. A general election is an instance of collective social action. Particularly where there are more than two realistic, alternative parties to vote for, casting a vote becomes subject to important strategic calculations about how other voters in your

constituency will act. This might be expected
to make for pronounced spatial effects since the
electoral unit is a geographical one. Electoral
systems constrain voters: in a national system
of proportional representation I would vote as
a citizen, in a syndicalist system I would vote
as a member of a functional group, but in the
British system I vote as a resident of a territorial
unit. The weight of my vote depends upon the
behaviour of my co-residents.

Class Voting: A Paradox

Ivor Crewe and colleagues on the British Election
Survey at Essex have established a dominant paradigm
for the understanding of recent British voting
behaviour (Crewe et al 1977). The central propositions of that paradigm are relatively well known:
voters are more volatile than they used to be;
loyalty to the two major parties has declined,
people are less strongly partisan than before;
class is no longer a powerful predictor of an individual's vote, indeed no social or individual
attribute or characteristic gives a good prediction
of electoral choice; voters seem to be voting on
the basis of their opinions about political issues
rather than on the basis of established loyalty
or group interest. The principal beneficiaries
of these changes have been 'third parties', the
principal loser, the Labour Party.
    Some elements of the partisan dealignment
thesis are more accurate than others. For present
purposes - an examination of the relationship between class, locality and voting behaviour - it
is necessary to consider only those parts of the
thesis dealing with class dealignment and partisanship. There can be no doubt that there is now
a poor statistical relationship between an individual's occupational class and electoral choice.
Since the mid-1960s the probability has been much
reduced that white collar workers will vote Conservative and blue collar workers Labour. This
is demonstrated convincingly by Butler and Stokes
(1974), Crewe, Sarlvik and Alt (1977), Sarlvik
and Crewe (1983) and Rose (1980).
    However, Miller has shown that if one examines
voting in constituencies, rather than the votes
of individuals, the class polarisation of the two-
party vote showed no tendency to fall in the 1960s
or 1970s (Miller 1977, 1978, 1979). If anything,
polarisation increased after 1966. The class

composition of constituency electorates was a constant and very effective predictor of voting. The importance of class, if measured properly, had not changed much from the 1920s to the 1970s. This finding presents an intriguing paradox which has received very little attention in the recent British literature. Class apparently becomes more important in the determination of constituency results whilst at the same time individuals are abandoning their erst-while tendency to vote on the basis of their occupational class. There are several possible explanations of this, but before moving onto these let me illustrate the enormous discrepancy between the conclusions based upon survey data and those based upon aggregate data.

Sarlvik and Crewe (1983) analyse the 1979 election results. On the basis of a survey of 1,893 respondents, using 'tree' analysis, they show that the variable which explains most variance in the vote in England is social class (manual/non-manual), which is just marginally better than either housing tenure or union membership (of someone in the family). However, none of these explains very much of the variance and, indeed, it is one of the arguments of the book that all statistically significant social variables taken together explain only half as much of the variance as do electors opinions about political issues (p.113). The 'explained proportion of the variance' of all social characteristics taken together for the two major parties is only 0.20 and 0.21 (p.370). Miller (1979), working on the same election in England, showed that class, when measured by the percentage of employers and managers among the occupied and retired males, explained almost four-fifths of variance in the two-party vote between constituencies ($R^2 = .78$). This was the same as in October 1974, and was notably greater than at the elections of 1964, 1966 and 1970.

Aspects of the Paradox

The differences between these two accounts may be examined under a number of headings. First, there is the question of the merits of individual, survey-based inquiry against aggregate data analysis, or, in other terms, of individualist and ecological fallacies. Second, there is a question of the meaning of the different measures of class used in the two accounts. Third there is the issue of the relationship between class and space, since recent voting behaviour, perhaps especially the 1983 elections, makes local and regional variation in partisanship seem very important.

Individualist and Ecological Explanations

There has been a long debate in international social scientific circles about the relative merits of individual and ecological analysis. In Britain, the ecological fallacy is more often exposed than the individualist fallacy (for a survey of the debate with reference to psephology see Dogan and Rokkan 1969). British election studies have been primarily survey based and have sought to explain an individual's voting on the basis of that individual's social characteristics. There has been an antipathy towards ecological analysis. This stems partly from problems of the data which is available. Unlike in many other countries, votes in British national elections are counted at the constituency rather than the ward level. Also, until 1966, census data was hard to adjust for political boundaries. Consequently, the areas for which ecological analysis can be undertaken are much larger than in other countries and the results tend to be less accurate and reliable. Nevertheless there also exists an unnecessary suspicion of ecological analysis. It is true that aggregate analysis makes it impossible, strictly speaking, to make causal statements about the behaviour of individuals. It is true that the excessive presence of any particular social group and an excessive tendency to vote for some particular party is not proof that this group is responsible for that excess voting. But the same could be said about any structural analysis or collective behaviour analysis.

Consider, for instance, Crewe (1973). In this article Crewe takes a sample of constituencies where there is a high level of manual worker car ownership, which he calls 'affluent worker constituencies', and a sample of mining constituencies, which he refers to as 'proletarian worker constituencies'. He then shows a propensity for high levels of Conservative voting in the former constituencies, and conversely of Labour voting in the latter group. He then worries inordinately about the fact that it might not have been the affluent workers who actually voted for the Conservative Party (indeed, other kinds of social processes might account for excessive Tory voting

- status panic on behalf of white-collar workers faced with affluent manual workers, for example). Crewe therefore asserts that aggregate data is only useful as description of a political context, unless other data about individuals can be advanced to prove that it was the affluent workers and not other people who were voting Conservative. This is unnecessary, unsociological, and perhaps over-fastidious, even from Crewe's own point of view. Individual behaviour is not the only object of sociological concern; to show that contexts have effects is valuable. If a high density of affluent workers always produces above average Tory voting then that is an interesting social fact, even though we may not be sure by what process those affluent workers interact to produce that effect. It can be argued that individual-level survey data are no more a guarantee of meaningful explanation than ecological analysis. If we find that it is the affluent workers who cast votes for the Tories we are still left with a problem of explanation as to why workers with cars should vote Tory - they surely won't tell you that that is the reason, which is the principal verificatory capacity of the survey method, where the respondent confirms directly the social scientists's hypothesis.

It is fairly widely acknowledged that ecological and individual level data do produce different conclusions (Dogan and Rokkan 1969, p.41). Miller's findings must warm the heart of Durkheimians: class appears to be a social fact which extrudes through the behaviour of individuals without correspondence with their intentions or motivations. However, as most advocates of ecological analysis would agree, there is a link between the two levels. It is worthwhile trying to explain what aspects of collective political behaviour produce Miller's paradox.

Measuring Social Class

It remains possible that the paradox derives from different definitions of class. Survey analysis has usually used a manual/non-manual distinction as a measure of class. In fact, Butler and Stokes's first edition of Political Change in Britain (1969) grouped routine white collar workers with manual workers when measuring 'objective class', though they reverted to manual/non-manual in the second edition(1). This is a measure of occupational status. It is a cleavage which has been important

in social and political life in Britain, but it
does not describe two social classes. The distinction is invariably one related to men. Male occupational status is measured and that status vicari-
ously attributed to married women (indeed, there
is a remarkable concatenation of categories which
obscures gender differences: occupational status
is a property of the husband, trade union membership
a property of the family, whilst housing tenure
is a property of the household) (see Peake in this
volume). Furthermore, it should be noted that
whilst occupational status is a declining predictor
of the individual's vote, other contending characteristics which are shown to explain variance (housing tenure, union membership, car ownership, etc.)
are aspects of material life. Finally, it should
be pointed out that different measures of declining
class polarisation are used in survey analysis.
Each produces somewhat different results, though
the trends they indicate are the same.

Miller uses a measure of class more fitting
to its object than that used in survey analysis.
He gives an extended critique of orthodox survey
analysis approaches to the attribution of occupational status to respondents (1978 pp. 258-264).
Instead, he uses the proportion of employers and
managers among occupied and retired men as his
measure of class (this is perhaps the only occupational measure which will not be confounded by
using a statistic for men alone, there being so
few women in such occupations that variation from
constituency to constituency will be slight). Moreover, 'employers and managers' does appear to define
a functional group with constant, homogeneous political tendencies, who bear a particular relation
to production and have a common set of political
and economic interests. In Electoral Dynamics
(1977), Miller showed that, using this measure,
there was a remarkable persistence in the salience
of class in explaining constituency voting behaviour. He shows that this measure (based on
social class II in the 1951 census, and obtainable
directly from the 1966 census) explains a vast
amount of variance, far more than any other, and
does not change during the 1970s (see also Miller
1979).

Space and Class

Any resolution of this paradox of class alignment
has to make reference to spatial effects.

If constituencies are voting on class lines but individuals are not, there must be some social processes occurring in localities which exaggerate partisanship in extension of direct class effects. This is particularly clear using Miller's measure of class, since clearly it cannot be the votes of managers and employers themselves which makes a difference, since they are too few in any constituency and in any case they are consistent Conservative supporters everywhere. Rather, their presence must be considered contagious. High local concentrations of managers and employers directly or indirectly affect members of other classes, causing them to prefer the Conservative Party to the Labour Party. Miller produces a figure for the excess of class voting by concentration of employers and managers for the 1966 election:

> the increments in Conservative support produced by one per cent more employers and managers in the locality, were 0.5% for those in the top class, 2.7% in the intermediate class, and 1.9% in the bottom class. (1978, p.278).

The existence of effects of this kind are far from unknown to psephologists and geographers of elections. They are referred to generally as 'neighbourhood effects'. Butler and Stokes (1974) demonstrated a blue collar/white collar neighbourhood effect, in that curvilinear relationship existed between the density of occupational status groups and the propensity to cross-class voting. A general tendency was identified whereby in constituencies where there was a high proportion of middle class people both the middle class and the working class vote for the Conservatives was greater than average, and vice versa for density of manual workers and Labour support. One of the clearest examples was their comparison of two-party support by class for mining constituencies and seaside resorts (see Table 1).

They suggested two possible explanations of this effect: (i) people perceive and conform to local political norms; (ii) informal contacts - on the shop floor, in pubs, in other face-to-face groups - are persuasive. They went no further than this and merely suggested that apparent variations in partisanship by region 'may reflect the summing up of numberless face-to-face contacts across the local areas that comprise different regions' (p.136). They do, however, demonstrate

41

Table 1: Partisan Self-Image by Class in Mining Seats and Resorts 1970

Mining Seats

	Class Self-Image	
	Middle Class	Working Class
Conservative	50%	b21%
Labour	c50%	79%
	100%	100%

b − c = −29%

Resorts

	Class Self-Image	
	Middle Class	Working Class
Conservative	80%	b52%
Labour	c20%	48%
	100%	100%

b − c = + 32%

Source: Butler and Stokes (1974, p. 131)

that regional variations in party support are the result not so much of variations in class structure as variations in class-party alignments in the regions. Between 1963 and 1966 manual workers in Wales and the North East preferred the Labour Party to a much higher degree than those in the South West and the South East (See Figure 2).

At present there is very little work available on how the neighbourhood effect operates. This is at least partly because it seemed to be a relatively unimportant effect to British psephologists working with survey-based data. Indeed, for reasons of sample size, no survey has ever been able to analyse regional, not to mention local, variations. Butler and Stokes (1974, pp. 121-130) and Rose (1980) have used a substitute strategy based on amalgamating the findings of regular opinion poll data (Butler and Stokes, for example, had a sample of 120,000 interviews, collected over three years, which they used to explore regional variation). Miller's ecological analysis (1977, 1978), however, makes the effect seem much more important than was previously imagined. This particular form of the neighbourhood effect isolated in aggregate data I shall call the 'class context effect'.

I shall first look at Miller's own explanation of the class context effect. I shall then subject it to criticism in the light of other empirical and theoretical work on spatial aspects of voting. Finally I suggest some lines of investigation to better understand that effect.

Miller's Explanation of Miller's Paradox

In an article, 'Social class and party choice in England: a new analysis' (1978), Miller used survey-based data collected by Butler and Stokes, and by Crewe, Sarlvik and Alt to try to reconcile his aggregate data with survey findings and thus resolve the paradox. He argues that the definition of class used in dealignment arguments is sub-optimal, and that the measure of class polarisation in the electorate is affected by that definition. He suggests that a more appropriate definition of class would show rather less of a decline in individual class alignment than that outlined by Crewe et al (1977). Nevertheless, he recognises that, with any measure, when one considers survey respondents, there has been class dealignment. Miller then considers a variety of ways in which neighbourhood effects might work, and in order to estimate

Figure 2: Political party support from social grades
by region, 1963-1966

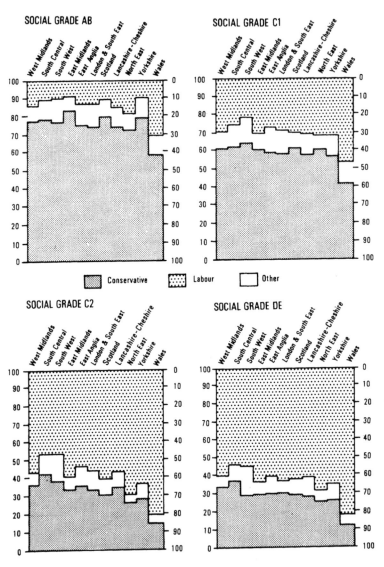

Source: from Butler and Stokes (1974, pp. 126-7)

their importance separates out individual class polarisation from environmental (constituency) class polarisation. His estimate shows that while the individual effect has fallen, the environmental effect has increased. In other words, the relative constancy of the predictive capacity of class at the constituency level, even using a manual/non-manual division, as Miller does in this instance, is accounted for by the fact that the environmental effect has increased to compensate for the decline in individual effects.

Miller interprets the environmental effect in terms of the concept of 'core classes'. For Miller there are two core classes, roughly speaking employers and managers on the one hand, and unionised manual workers on the other. The political predilections of these two classes are predominantly fixed, and the members of those classes will vote Conservative and Labour respectively without regard to spatial or contextual factors. All other voters are, however, responsive to class environment, where class environment is defined by the local ratio of core classes. This assumes that class is more salient to members of core classes than to other people. Indeed, Miller effectively seems to say that there are only two classes, to which very few people belong. Most people are merely linked to social classes. Given that Miller uses a powerful notion of class compared with the manual/non-manual division, his analysis would seem capable of explaining the more pervasive neighbourhood effects suggested by his ecological analysis in Electoral Dynamics (1977). He goes on to show that using the core class of 'employers and managers', class polarisation at the constituency level increased between 1964 and 1974, and a numerical estimate can be given (as quoted above) of the influence of the presence of that core class on other occupational classes.

Miller also comments on the significance of this increasing class context effect, and the fall of the individual effect. He contrasts England with Northern Ireland where individual effects (of religion, of course) explain almost all the variance in voting (in the 1960s 99.5 per cent of Protestants voted for unionist parties, while only 5 per cent of Catholics did so). The extent of contextual partisanship in Great Britain, by contrast, may be, says Miller, evidence of a well-integrated, tolerant and consensual electorate. He also noted that the context effect, whereby a leading core class can mobilise other groups locally, means that area-based forms of political partisanship and action are likely to become more important. 'When the class alignment becomes less and less about "people like me", more and more about "people around here" it should not be altogether surprising if there is increasing scope for other, more class-less responses by "people around here" ' (1978, p. 283). He cites the success of the Scottish Nationalist Party in the 1974 elections as an instance of such an emerging local/regional politics.

Discussion

Miller's work is extremely interesting. Many of the suggestions he makes seem to be supported by recent events and other kinds of analysis of electoral change. Yet his work is almost entirely neglected. I want both to suggest some ways in which his work is corroborated by other material, to extrapolate some of his findings, to criticise and sharpen some of the concepts, and finally to suggest ways in which his hypotheses might be fruitfully tested.

Local Consensual Partisanship

Miller suggests that constituencies will tend to develop into partisan localities, subject only to the limit that members of both core classes are unaffected by context. This suggests how it can be that, given high rates of individual electoral volatility, there is still such a thing as a safe seat. A recent analysis by Curtice and Steed (1982) supports Miller, showing that since 1955 voting for the two major parties has been subject to persistent spatial trends along two dimensions, urban/rural and north of Britain/south of Britain. Labour has increased its hold over urbanised and northern constituencies; the Conservatives have done the same in less urbanised and southern constituencies. One consequence of this was that the number of marginal seats in Britain has steadily declined.

Curtice and Steed show very persistent cumulative variations in the swing in different types of constituency. Table 2 shows that the swing varied along the dimensions urban/rural and north/south of Great Britain (Northern Ireland is excluded). This table shows that there were two

Table 2: Mean Swing, 1955-79, Broken Down by
Region and Rurality

	South & Midlands	North of England	Wales	Scotland	Mean
	(a) 1955-70: overall (United Kingdom) swing +0.09				
City	-1.93	-7.51	-	-9.53	-3.92(113)
Very Urban	+0.08	-5.60	-4.55	-9.02	-2.78(110)
Mainly Urban	+1.19	-2.09	-4.47	-6.81	-1.08(123)
Mixed	+3.01	-0.22	-4.16	-5.74	+0.40(147)
Mainly Rural	+3.59	-1.35	+5.16*	-2.06	+2.17*(77)
Very Rural	+4.97	+1.29	+0.42	-1.40†	+1.68†(36)
Mean	+1.21	-2.98	+2.89*	-5.82†	-0.98
	(335)	(171)	(31)	(69)	(606)
	(b) 1970(NT)-79: overall (United Kingdom) swing +2.44				
City	+0.24	-4.74	-	-8.12	-1.98 (99)
Very Urban	+1.48	-1.81	+1.45	-7.15	-0.28(100)
Mainly Urban	+4.26	-0.60	+2.42	-4.22	+1.67(144)
Mixed	+6.24	+2.77	+6.17	-2.17	+4.48(158)
Mainly Rural	+7.87	+3.75	+10.49	+0.31	+6.27 (76)
Very Rural	+8.82	+7.99	+15.16	+4.03	+8.45 (40)
Mean	+4.24	+0.10	+7.60	-2.84	+2.49
	(344)	(170)	(34)	(69)	(617)

Definitions:
South & Midlands: five southernmost standard regions in England as defined by the Registrar General before 1974; North of England: three northernmost standard regions in England; City: electorate density greater than 35 persons per hectare in 1965; Very Urban: electorate density greater than 22 persons per hectare and less than 35 persons per hectare; Mainly Urban: electorate density greater than 6 persons per hectare and less than 22 persons per hectare. Mixed: electorate density less than 6 persons per hectare and proportion of the employed population engaged in agriculture less than 6 per cent; Mainly Rural: proportion engaged in agriculture between 6 and 15 per cent; Very Rural: proportion engaged in agriculture 15 per cent or more.
    For the 1955-70 constituencies the proportion of the employed population engaged in agriculture was obtained from the 1966 Census; for the 1970(NT)-79 constituencies from the 1971 Census. This inconsistency had to be accepted because attempts by the authors to obtain a copy of the 1971 Census Economic Activity Data for 1955-70 parliamentary constituencies were unsuccessful, a copy seemingly not having been retained either by OPCS or by anyone in the academic community. OPCS are unable to reprocess the 1971 Census until after completion of their work on the 1981 Census, and to wait for this would place an inordinate delay on the publication of our article.
    The census figures of those engaged in agriculture also include those engaged in fishing. Two constituencies where this pushes the figure above 6 per cent - Grimsby (both periods) and Hull West (1955-70 only) - are classified according to their electorate density rather than as Mainly Rural.
    The sources used in the calculation of the electorate density figures are detailed in fn. 30. Two-party swing is used in this table (see fn. 27). Seats where there was a split Conservative or Labour candidature, or an absence of a Labour or Conservative candidate in one of the elections concerned, are omitted in the calculation of mean swing.
*   These figures are heavily influenced by a very strong, untypical swing in Anglesey (viz. +20.52); if Anglesey is excluded, the following figures apply: -2.52 (Wales, Mainly Rural); +1.90 (Mean, Mainly Rural); -3.67 (Wales, Mean).
†   These figures are similarly influenced by a strong swing in Caithness & Sutherland (-28.26); if Caithness & Sutherland is excluded, the following figures apply: +0.84 (Scotland, Very Rural); +2.53 (Mean, Very Rural); -5.49 (Scotland, Mean).

Source: Curtice and Steed (1982) Table 1, p. 257

phases in the process, 1955-70 and 1970-79; that there were regional variations; and that there was a significant overall urban/rural dimension to swings. Because party support is becoming more concentrated, the number of marginal seats has declined. Marginal seats were defined (prior to the 1983 election) as those in which the share of the two-party vote, for either major party, lay between 45 per cent and 55 per cent. This figure was taken because the proportion of the two-party vote taken nationally at all elections since 1945 had fallen within those limits. Any constituency where one major party had more than 55 per cent of the vote would be a safe seat. On this basis they showed that the number of marginal seats fell steadily, with one exception, between 1955 and 1979 (see Table 3). The implication is that the major parties are increasing their grip on those constituencies which they usually won, and that fewer and fewer seats would change hands between elections. Curtice and Steed predicted the continuance of this trend.

Table 3:

Year	Standard deviation	Kurtosis	No. of marginal seats	(N)
1955	13.47	-0.25	166	(610)
1959	13.79	-0.29	157	(610)
1964	14.12	-0.45	166	(609)
1966	13.77	-0.46	155	(605)
1970	14.34	-0.27	149	(608)
1970(NT)	14.32	-0.30	149	(614)
1974(F)	16.10	-0.68	119	(598)
1974(O)	16.82	-0.82	98	(596)
1979	18.86	-0.87	108	(607)

* The table is based on those seats fought by both Conservative and Labour and won by either of them. The kurtosis given is the difference between the kurtosis of a normal distribution and that of the actual distribution. For definition of marginal seats see text, p.269.

Source: Curtice and Steed (1982), Table 2, p. 268.

Curtice and Steed's projections seem consistent with the analysis of Miller. Both spatial trends, which seemed to be accentuated in 1983, especially with respect to Labour Party support, are consistent with the class contextual effect. In advance of full statistical analysis of the 1983 election I cannot be sure, but it would seem that the geographical concentration of Labour support in the conurbations and in the older industrial regions of the north of England, central Scotland and South Wales, might be consistent with Miller's hypotheses about the effects of core classes. These might be the only areas where there was a core class ratio favourable to Labour. In such areas Labour support remained firm. In other suburban and rural areas Labour support was more or less obliterated. The consequence of this is of course that survey based responses will show a remarkable fall in individual class polarisation: since so few people in the south voted Labour, national aggregate figures will show a very weak class-party alignment for individual manual workers voting Labour. The class contextual effect may however remain. McAllister and Rose's aggregate data study of the 1983 election, The Nationwide Competition for Votes (1984), shows that the socio-economic status of a constituency explained most variance in electoral outcomes for Conservative, Labour and Alliance parties. Territorial factors were also significant. There were distinctive 'national' differences between England, Northern Ireland, Scotland and Wales, but also intra-national differences too. Region and distance from the national capital also 'explained' substantial amounts of variance (see Table 4 for a summary of social and territorial influences on the 1983 vote in England).

Core Classes

The concept of core classes derives from the work of Richard Rose (1974). He used it to describe two ideal-typical individuals, an archetypal middle class person and an archetypal working class person. The worker was in manual employment, a member of a trade union, lived in a council house, had a minimum education and identified subjectively with the working class. The middle class ideal type was, conversely, in non-manual work, not a member of a union, was an owner-occupier, has some post-minimum education and identified with the middle class. Rose introduced this concept in order to

49

Table 4:

	Conservative vote		Labour vote		Alliance vote	
	b	% variance explained	b	% variance explained	b	% variance explained
Social Structure						
Socio-econ status	.37**	42	-.46**	39	.08**	9
Agriculture	.11**	15	-.16**	17	.06**	8
Immigrants	-.02	2	.04*	4	-.01	2
Elderly	.01	1	-.04*	3	.02	2
Territory						
Miles from London	-.03**	11	.01	4	.00	1
Region^a	2.8**	7	6.9**	12	3.4**	8
	—	—	—	—	—	—
(Constant) r^2	(24.2)	78	(55.3)	79	(18.1)	30

a   South of England for Conservatives and Alliance; Labour, North of England.

*   Significant at .05 level

**   Significant at .01 level

Source: I. McAllister and R. Rose, Table V.1, p.89.

reconcile the fact that the colour of a worker's collar was a relatively poor indicator of partisanship, but that many of the variables which were equally, or even more, important in predicting votes (housing, union membership, etc.) seemed to be functions of a person's role in the division of labour.

Survey research does show that the more of such ideal typical traits a person has the more likely they are to vote for their class party. Sarlvik and Crewe (1983), for example, show that in the 1979 election, of the three party vote,

71 per cent of manual workers, who were council tenants with a personal or family affiliation to the TUC, voted Labour. Seventy-three per cent of non-manual, non-TUC affiliated home-owners voted Conservative (these groups comprised respectively 13 per cent and 29 per cent of voters). Rose points out that by 1979 only 14 per cent of the electorate conformed exactly to either of these class stereotypes (Rose 1980).

Miller's use of the term is in some respects more, in others less, demanding and precise. All his aggregate analysis is conducted using the core middle class (i.e. employers and managers). As he points out, only this group can be identified from census data. Sociologically we might, a priori, assume this to be a meaningful definition or delimitation of a social class. The fact that Miller obtains stronger environmental class effects using this definition than do Crewe and Payne (1976) or Johnston (1981), both of whome use the percentage of workers in manual occupations as the index of class in ecological regressions, suggests that his is a class category with extensive causal powers. The existence of the working 'core class', however, is merely an hypothesis. Miller seems to consider it as comprising unionised manual workers, though he says that its boundaries may be relatively indefinite. The fact that we have little notion of the spatial distribution of union membership makes quantitative estimation of its local presence impossible. Hence, though Miller's reasoning seems highly plausible, there are severe obstacles to any precise test of the importance of the core class ratio on local partisanship. Nevertheless, it is appealing to imagine that we might obtain a systematic quantifiable index of the intensity of class conflict!

Miller, however, is able to make some statistical correlations between the presence of the 'controllers' (i.e. managers and employers) and the behaviour of other classes[2]. The figures I quoted above estimate the average effect of 'controller' presence on other occupational classes. Miller was also able to show that the behaviour of intermediate occupational groups is sensitive to the density of 'controllers'. People in contradictory class locations (Wright 1976) are most sensitive to class environment (Miller 1977, pp. 58-60). Thus, Miller shows that the bourgeoisie exercises local partisan 'leadership' on behalf of the Conservatives in England when present in

51

comparatively large numbers, and that its relative absence encourages Labour partisanship. However, there is no way of determining whether it is a general antipathy to the 'controllers' or the positively orchestrated 'alternative hegemonic' role of a 'working core class' which accounts for Labour partisanship. It must not automatically be assumed that labour mobilises in the same way as capital. Indeed, Offe and Wiesenthal (1980), among others, provide strong reasons against assuming symmetry in class organisations.

There is one notable omission in Miller's account of the role of core classes. He fails to specify how core classes produce the effects attributed to them. A whole range of mechanisms could operate to cause the effect of local political hegemony. It is unfortunate that Miller pays no attention to this question. It does however present the possibility of empirical research testing to corroborate or qualify an eminently plausible hypothesis. Are core classes responsible for neighbourhood effects?

The Neighbourhood Effect

Explanations of the neighbourhood effect are pretty sparse. The authoritative piece of empirical research on the subject was completed by Putnam and published as long ago as 1966. He noted a tendency for local communities to develop political traditions, so that they maintained a party preference over very long periods of time, apparently transmitting that preference through generations and across various social classes. He reasoned that there were three possible explanations of the concentration of partisanship in local communities. Each could account for Butler and Stokes' findings. These potential explanations are: (i) parties may be more active in areas where they are strong and persuade more people to vote for them; (ii) there may be perceived community political norms and people tend to conform to those norms; and (iii) the neighbourhood effect may be a function of social interaction, such that people persuade their contacts, their friends and people whom they meet in local associations, to vote in the same way. Putnam contrived an experiment based on two steps. First he separated out United States counties with different degrees of support for the Democratic candidate in the 1952 and 1960 presidential elections. He then interviewed to explore the three hypotheses. He inquired whether parties were more active where they were strong. He tried to discover whether people with attachments to the locality perceived the existence of a community norm, or felt any compulsion to act in accordance with the dominant pattern in the community. Further, he examined whether high degrees of social interaction increased the likelihood of voting in accordance with the community pattern. Though all were relatively difficult to measure, he found no confirmation for the first or second hypotheses. The third however seemed to be significant. In particular, people who were members of formal associations in the locality were much more likely to vote in accordance with the community norm, whatever that happened to be. Membership increased a person's integration into the community and, Putnam argued, brought him or her into contact with a broad cross-section of the community. If you attend public association meetings, whether parent-teacher associations, a women's institute, or whatever, the partisan complexion of those bodies is likely to represent the balance of partisanship in the community. In this way, partisan communities are likely to become more partisan. This is of course different from talking to one's friends, for informal groups are likely to be formed among people of like-mind anyway. Even then, in a strongly partisan community it is statistically difficult for dissenters to find people of like-mind to consort with: finding a group of Labour voting friends in the Fylde may be difficult.

It seems to me that this kind of explanation may go some way towards explaining Miller's and Curtice and Steed's (1982) findings. Those constituencies or localities which are strongly partisan are likely to see an increase in partisan support locally if there is considerable vitality in civil associations in that locality. We know there are considerable class differences, and we would suspect that there are considerable gender differences, in associational membership. We would expect that there would be a very strong representation of certain kinds of people in these organisations. The enhanced strength of Tory support might then be expected in Conservative localities in the south and in rural areas. In Labour strongholds it presumably is a matter of the degree of participation of working (core) class people in such associations whether they continue to define a majority community pattern of partisanship. It is then a matter of some importance to ascertain to what extent working class and Labour-supporting professionals engage in civil associations.

Information concerning the membership of civil associations by locality in Britain is limited. Arguments about privatisation and about the decline of working class politics would suggest that working class participation would be minimal. If this were so, and Putnam's conclusions hold for Britain post-1960, it would be difficult to account for the consolidated Labour support reported by Curtice and Steed (1982). In fact there probably remains quite extensive working class membership of civil associations. Most obviously, trade union membership increased steadily until the beginning of the 1980s. The size and range of membership of voluntary associations recorded in Social Trends 1983 is remarkable, though I am not sure quite what partisan effects I expect as a result of knowing that there are just less than a million members of the British Legion! Goldthorpe et al (1980) reported that, in the Nuffield social mobility sample, the mean membership of clubs and associations was 1.5 for inter-generationally stable manual workers, compared to 3.6 for inter-generationally stable members of his class I. It seems likely then that working class membership of civil associations is quite extensive. It is also likely that it will be spatially differentiated. The presence of working men's clubs has made a difference to neighbourhood partisanship in Preston in the inter-war period (Savage, 1984), and this may be part of the explanation of continuing Labour support in the North. Anecdotally, Michael Meadowcroft, elected Liberal MP for Leeds in 1983, in part attributed his success to the existence of some half a dozen Liberal social clubs in the constituency in which people could be addressed as if they might be Liberal partisans (Meadowcroft, 1984). Also, intuitively, one might expect working class participation in suburban areas to be less extensive, also perhaps in areas of high population turnover (new towns, etc.).

Class and Status Group Politics

One of the interesting features of Miller's analysis is that he predicts the emergence of classless political alignments in localities on the basis of environmental class polarisation. People take a view about what is in the interests of 'people around here' on the basis of what might be called local class hegemony. This might be interpreted as a peculiar consequence of the institutional- isation of class conflict. Structural class an- tagonisms are negated by local concentration of core class members who, somehow, align the com- munity behind them. This suggests an alternative explanation of the basis of Celtic nationalism, of regional political differences, and of local social movements. In many such accounts class politics is directly opposed to status group, ethnic group, or local politics. Miller suggests that rather than being mutually exclusive there is a necessary connection between class and other kinds of movements. Such a suggestion is corroborated by Ragin (1977) with respect to the bases of electoral behaviour in Scotland and Wales.

One of the most substantial and ambitious works of aggregate data analysis in British politics was Hechter's Internal Colonialism (1975). Critics, whilst finding it fruitful, have shown that work to be seriously defective in many respects. Ragin reworked Hechter's material to very good effect. Hechter had maintained an 'internal colonialist' explanation of the relationship between the Celtic periphery and the English centre. Economic under- development of the periphery had engendered a cul- tural division of labour within the UK. Something akin to dominant and subordinate ethnic groups had been created. These quasi-ethnic cleavages were manifested in voting behaviour, as 'peripheral sectionalism'. Hechter argued that class or func- tional divisions could not explain peripheral sectionalism and that an alternative approach in terms of 'status group politics' was required.

Hechter's work was presented as a critique of a 'modernisation' (or 'developmental') account of political change. Such an account was deemed defective because, among other reasons, it assumed that in the process of development status group differences would give way in functional cleavages. Ragin argues that neither approach is correct, though 'the contradiction between these two argu- ments is more apparent than real' because 'Celtic sectionalism has roots in both class and status group affinity' (1977, p. 439). Drawing on Wallerstein (1974), Ragin argues that it is class conflict in peripheral areas which encourages the formation of culturally-based peripheral opposition to core areas, noting that the origins of that sectionalism lie in local dominant strata. Ragin shows that support for Labour in the periphery is explained by class alone, but that there is an anti-Conservative effect among intermediate and dominant strata.

Ragin shows, contrary to Hechter, that the Labour Party never benefited from peripheral sectionalism. Labour support between 1924 and 1966 was spatially undifferentiated; the density of manual workers by county in the Celtic periphery was sufficient to explain Labour support. There were no systematic sectional effects. In this respect, Labour support was consistent with the modernisation theory account. Hechter, however, had taken Conservative support as his dependent variable. There were peripheral sectional effects in this case; Scotland and Wales were systematically anti-Conservative. That proclivity was actually, though, expressed through unusual degrees of support for the Liberal Party, and later, the nationalist parties. Neither of the main proponents of the modernisation thesis, which asserted the disappearance of status politics and a tendency to inter-regional uniformity had picked up on this however; Alford (1963) because he had used Labour support as his dependent variable, Butler and Stokes because they only considered support for the two major parties. The moral of that story is that party support is asymmetrical with respect to social bases.

The importance of Ragin's account, here, is 'that the rise of class-based political action does not imply a concomitant decline in political regionalism' (p. 448). Rather, Ragin says, peripheral sectionalism continues after the development of class politics, 'but the strength of the peripheral response is limited by class factors' (p.448). In this case Ragin was thinking of Labour's support in the periphery. He also contends that his findings support the Wallerstein hypothesis that cultural or status-based peripheral opposition emerges among the local dominant strata, and that 'such opposition is arrested, or at least limited by class cleavages. Thus, emphasis on local culture in peripheral areas may be based on both class and status factors: it cannot be explained simply by one or the other' (p.449). Class cleavages which emerge with industrialisation in the periphery hinder the extension of status group affinity (hence, Ragin suggests that the absence of class politics in Ireland was a result of industrial development being weaker).

Ragin appears to give a plausible account of the relationship between social bases and electoral behaviour in the Celtic periphery. As such it is consistent with some of Miller's observations on the differences in Wales and Scotland, and the lesser contribution made to electoral results by the presence of employers and managers in Scotland (Miller, 1979). More importantly, perhaps, the notion of the survival of status group politics, confined by class cleavages, seems to throw light on regional movements of all kinds.

In these accounts the concept of status group politics is to a degree specious. It says little more than that there are systematic collective effects not directly attributable to class structure. Most aggregate analyses have found such effects which they are usually unable to explain. For instance, McAllister and Rose (1984) identify 'territorial factors', and Crewe and Payne (1976) uncover 'past partisanship' as accounting for variance between constituencies, but both interpret these as entirely independent of class. At least in principle, the idea of class constraints upon such collective effects, as proposed by Ragin, might help to resolve Miller's paradox.

It was a weakness of Miller's notion of a core class ratio that he fails to predict the political consequences of specific values of the ratio: it being unlikely that electoral effects are in a linear relationship with the actual values of that ratio. It is worth speculating whether there is a point, when the ratio is heavily in favour of the 'controllers', where constituency alignment on the basis of class ceases. If Labour support can be accepted as an expression of core working class interest, it might seem that the 1983 election results of suburban and rural areas in southern Britain bore no relation to divisions of class interest. Should Alliance support in such places be interpreted as an instance of status group alignment?

Some Conclusions and Suggestions for Further Inquiry

Miller's paradox remains substantively unexplained. The model he offers is shown over.

He maintains that these effects occur simultaneously, but that, in England especially, the environmental component has become stronger since 1966, while the individual component has weakened. As a result, until 1979, the overall class effect

CLASS EFFECTS ON ELECTORAL OUTCOMES

Environmental component
core class ratio $\longrightarrow$ political $\longrightarrow$ exaggerated
environment   local
partisan
effects

Individual component
occupational $\longrightarrow$ instrumen- $\longrightarrow$ voting
position   tally   choice
rational
calculation

remained constant. The principal problems with
this account are threefold. First, there is no
explanation of why in the 1960s there is a change
in the degree of determination of the individual
and environmental components. This seems to require
an explanation both of what changed in party politics which transformed voters' responses, and
why there should have been such powerful individual
class effects in a period of consensus politics.
Second, it requires better specification of the
character of the core class ratio. The role of
the working core class is unspecified, partly
because it is unmeasured. There is a need for
precise specification of the electoral effects
when the ratio has different values. The parameters
of the ratio need to be explicitly stated. Third,
there is no sociological explanation of how the
core class ratio generates the environmental class
effect. We need to know a great deal more about
political communication in neighbourhoods and constituencies, about inter- and intra-household political communication processes.
   Towards the latter end, I tentatively suggest
that we might operate with some notion of local
political hegemony. The kind of account that Parkin
(1967) gave of working class conservatism
is interesting in this respect. He stood on its
head the problem, which obsessed psephology in
the 1960s, of why workers voted against their class
interests. He argued that all people would 'normally' or 'naturally' vote Conservative because
of the power of the dominant value system. Only
where there were strong barriers to that value
system, barriers created by working class communities and occupational solidarities, would there
be significant degrees of Labour voting. Parkin's
account contains two dubious assumptions: (i) that

there is a dominant, national value system which predisposes people to vote Conservative; and (ii) that there is <u>any</u> 'normal' or 'natural' voting response. These two assumptions could be abandoned for a conception of the way in which, locally, class structure produces hegemonic (i.e. persuasive and leading) classes which, within certain specified parameters (of the core class ratio), produce excessive environmental partisanship. This appears consistent with, and a more plausible explanation of, Miller's findings. It also has the interesting analytic consequence of connecting local (community) power structure with electoral behaviour.

The study of local political communication may be of considerable interest. It may make it possible to understand women's political behaviour better. The available evidence on women's voting is negligible, the interpretations of it arbitrary. Women are, presumably, affected in similar ways to men by the environmental component of class voting, but significantly differently affected by the 'individual component'. Also, though, women's participation in the labour market must partly create the 'environment'. Sociological inquiry into political communication - focusing on associational membership, household and neighbourhood connections (especially in the context of the de-industrialisation thesis), the desertion of local politics by mobile capital, the ambivalence of professional classes toward partisanship, the social bases of local social movements and 'status group mobilisation', etc. - appears to offer interesting possibilities. Such work would be concerned with the insterstices of the ecological and the individual effects which Miller brought to light.

To resolve Miller's paradox would, it seems to me, enhance considerably our understanding of the relationship between space, class and politics.

NOTES

1. The recent volume of D. Robertson (1984) argues that if a three class model (somewhat like that of the Nuffield Mobility Studies) is used as a basis of analysis, then class remains statistically very important in explaining the votes of individuals in the 1979 election.
2. For the concept of 'controllers' I am indebted to Rev. W. Awdry.

REFERENCES

Alford, R. (1963) Party and Society: the Anglo-American Democracies, Greenwood, Westport, Conneticut.

Butler, D. and Sloman, A. (1980) British Political Facts, 1900-1979, Macmillan, London.

Butler, D. and Stokes, D. (1969) Political Change in Britain: Forces Shaping Electoral Choice, Macmillan, London.

Butler, D. and Stokes, D. (1974) Political Change in Britain: The Evolution of Electoral Choice, 2nd. ed., Macmillan, London.

Crewe, I. (1973) 'The politics of "affluent" and "traditional" workers in Britain: An aggregate data analysis', British Journal of Political Science, 3, 29-52.

Crewe, I. and Payne, C. (1976) 'Another game with nature: an ecological regression model of the British two-party vote ratio in 1970', British Journal of Political Science, 6, 43-81.

Crewe, I., Sarlvik, B. and Alt, J. (1977) 'Partisan de-alignment in Britain, 1964-74', British Journal of Political Science, 7, 129-190.

Curtice, J. and Steed, M. (1982) 'Electoral choice and the production of government: the changing operation of the electoral system in the U.K. since 1955', British Journal of Political Science, 12, 249-298.

Dogan, M. and Rokkan, S. (eds.) (1969) Quantitative Ecological Analysis in the Social Sciences, M.I.T. Press, London.

Goldthorpe, J., Llewellyn, C. and Payne, C. (1980) Social Mobility and Class Structure in Modern Britain, Clarendon Press, Oxford.

Hechter, M. (1975) Internal Colonialism: the Celtic fringe in British national development 1536-1966, Routledge and Kegan Paul.

Johnston, R.J. (1981) 'Regional variation in British voting trends 1966-79: tests of an ecological model', Regional Studies, 15, 23-32.

McAllister, I. and Rose, R. (1984) The Nationwide Competition for Votes: The 1983 British Election, Frances Pinter, London.

Meadowcroft, M. (1984) 'The new (liberal) left: an interview', Marxism Today, February, 14-18.

Miller, W.L. (1977) Electoral Dynamics in Britain Since 1918, Macmillan, London.

Miller, W.L. (1978) 'Social class and party choice in England: a new anlaysis', British Journal of Political Science, 8, 257-284.

Miller, W.L. (1979) 'Class, region and strata at the British election of 1979', Parliamentary Affairs, 32, 376-382.

Offe, C. and Wiesenthal, H. (1980) 'Two logics of collective action: theoretical notes on social class and organisational form' in M. Zeitlin (ed.), Political Power and Social Theory, JAJ Press, Greenwich, Conneticut, pp. 67-115.

Parkin, F. (1967) 'Working-class Conservatives: a theory of political deviance', British Journal of Sociology, 18, 278-290.

Putnam, R. (1966) 'Political attitudes and the local community', American Political Science Review, 60, 640-654.

Ragin, C. (1977) 'Class, status and "reactive ethnic cleavages": the social bases of political regionalism', American Sociological Review, 42, 438-450.

Robertson, D. (1984) Class and the British Electorate, Blackwell, Oxford.

Rose, R. (ed.) (1974) Electoral Behaviour, Free Press, New York.

Rose, R. (1980) Class Does Not Equal Party, University of Strathclyde, Studies in Public Policy 74.

Rose, R. and Unwin, D. (1975) Regional Differentiation and Political Unity in Western Nations, Sage, Beverly Hills.

Sarlvik, B. and Crewe, I. (1983) Decade of Dealignment: The Conservative Victory of 1979 and Electoral Trends in the 1970s, Cambridge University Press, Cambridge.

Savage, M. (1984) Social Structure and Working Class Politics in Preston 1890-1940, unpublished Ph.D. thesis, Department of Sociology, University of Lancaster.

Wallerstein, I. (1974) The Modern World-System: capitalist agriculture and the origins of the European world economy in the 16th century, Academic Press, London.

Wright, E.O. (1976) 'Class boundaries in advanced capitalist societies', New Left Review, 98, 3-42.

CHAPTER FOUR

A CONCEPTUAL INQUIRY INTO URBAN POLITICS AND GENDER

LINDA PEAKE

Introduction

The intention of this paper is to raise some questions in relation to the following: the conceptual
categories that have been employed in research
on urban politics; the feminist imperatives which
require the redefinition of these categories; and
the implications of engaging in urban political
research. In doing so I want to suggest an alternative approach to the study of urban politics which
will challenge its existing parameters. Integral
to this approach is a demand for a clarification
of the object of study of urban politics. This
can be brought about by the admission, into its
frames of reference, of two simple factors; first,
that the female half of the population is as worthy
of study as the male half; secondly, that gender
relations are important elements of social organisation and change. The rendering of women and
gender relations as non-political has led to a
distortion of our understanding of urban politics
which needs to be rectified. This can be ac-
complished by adopting an epistemology which sees
questions of meaning and conceptual inquiry as
fundamental to theoretical development. The basic
premise of my argument is that both the meanings
attached to the categories 'women' and 'urban
politics' and the methodological approaches used
in their study need to be critically examined and
reconceptualised. This will hopefully enable us
to increase our understanding of both the meanings
that engagement in urban political activities have
for women and of the specific ways in which they
participate.

Urban Politics

Structural transformations of the British economy have consequences not only for economic performance but also for social and political behaviour and as such cannot be ignored in any analysis of the determinants of urban political practice. Concomitantly, theoretical developments in urban politics have arisen in response to activities taking place in the political arena. Indeed, the emergence of the new radical school of urban and regional studies was premised upon the realisation that new forms of social organisation and change were evolving in western capitalist societies (Lebas 1983). Counterposed to production and class struggle, consumption issues and urban social movements have increasingly become the object of political analysis with definitions of urban politics focusing upon the area of decision-making relating to provision of consumption goods and services by the state (Castells 1978, Dunleavy 1980).

In the post-war period the state has played an increasingly important role in the provision of housing, transport, education, health care and the like, which themselves have become increasingly necessary (although not sufficient) for the re-production of labour power on both a daily and generational basis. Consequently, the sphere of consumption has become increasingly collectivised (i.e. managed by state agencies with access determined by non-market criteria) with commodity production and capitalised services replacing many areas of domestic service and manufacture (Mackenzie, 1980). This socialisation of consumption comes to dominate patterns of daily life, exacerbating market-based inequalities. But as more aspects of daily life become obvious arenas of political regulation, they also become areas of political struggle; increasing socialisation has led to politicisation (Habermas, 1976). In particular Dunleavy has argued that interests in consumption are structured along a cleavage based on an 'individualised-commodity-private mode and a collective-service-public mode' and that an individual citizen's political allegiences, both in electoral politics and in locality-based conflicts, are influenced by their consumption locations (Dunleavy 1979, p. 419).

The preceding sketch reveals that the study of urban politics has centred upon structures and practices with an institutional basis tending to have an inherent male bias, from which women are usually absent. It has concerned itself with the exercise of public power, which is the sphere of male domination and ignored the system of power relations within the family. In addition, the 'natural' context of urban politics has been the state, which precludes the more specific context of women's participation. Finally, the concern with collective consumption as a mechanism for the social organisation of space abrogates itself of responsibility for the 'isolation and colonisation of women within the home and the notion of collective consumption as a basic expression of women the primary consumers' (Ettore 1978, p. 501).

My intention is to examine these gender-blind research practices, which, in giving rise to an inadequate understanding of women's capacities and interests have seriously undermined the study of urban politics and allowed the false assumption to be drawn that women are 'non-political': women's political experience - as all experience - has been devalued(1). An essential requirement of an alternative interpretation of women's role in urban politics is the deposing of its androcentric basis i.e. the 'male-as-norm' principle, by attributing new meanings to variations in women's participation in traditional areas of urban political activities and by re-evaluating women's contribution as 'private' persons to urban political activities. As Bonder (1983 p. 571) asserts, 'women perceive and practise political activity in conditions and in specific ways which do not coincide with the conception and practice of political activity based on the male pattern'. We need to determine the underlying factors which account for these differences and assess their implications for the development of urban politics.

The first problem we encounter is simply uncovering the position of women in relation to the political system (see Bondi and Peake, forthcoming). This is not an easy task as women have been, in effect, doubly excluded: as paid workers in the system of production they are treated as peripheral to the formation of social relations, and once relegated to the family sphere they are naturalised as consumers and treated as being simply 'there' i.e. as being beyond the bounds of the 'political', and are again excluded from the study of urban politics. Moreover, the relationship between women's participation in urban political activities and the primacy of their responsibility for work in the sphere of reproduction has not been seen as problematic (Mackenzie 1979).

The following section will address the feminist literature which accounts for the specificity of women's political position through an analysis of the changing relationship between the domestic and market economies i.e. the spheres of reproduction and production. Further to this an attempt is made to reveal how the site of the political arena also changes in response to definitions of what is political (Lovenduski 1981). 'Politics' is not the perogative of the public sphere, nor are women only located in the private sphere. It is the analytical distinction which has been drawn between the public world of employment and politics and the private world of the family and interpersonal relationships which allows men's experience to be equated with the public (political) sphere and women's with the private (apolitical) sphere. A reconceptualisation of the boundary between the public and private spheres reveals the opportunity for a reconstitution of the 'political' and is one of the first steps towards developing an urban politics that will address women's interests, needs and opportunities(2).

A Conceptual Inquiry

Much of the present feminist work within geography has concentrated upon establishing links between theory and practice with regard to two areas of study: manifestations of women's oppression and the relationship between gender inequality and modes of production (Foord 1983). Within the specific field of urban politics feminist research has sought to explain the specificity of women's social position through an analysis of the dialectical relationship between reproduction and the social relations of production, combined with an attempt to integrate this analysis into political practice (See Brownill n.d.; Mackenzie 1980; Bondi and Peake forthcoming). With this theoretical development the role of gender relations in urban politics has been made explicit. Traditionally, however, gender relations have been rendered invisible. Their neglect follows, in part, from the preoccupation with the (insufficient) concept of collective consumption: a sole concern with the (collective) means of the reproduction of labour power has been corresponded with a broader notion of the reproduction of the 'general conditions of production'. This unsatisfactory situation has arisen from 'an astonishing play on the word "reproduction" ' (Hindess and Hirst, 1977, quoted in Barrett, 1980) which not only obscures the conflation between biological and social aspects of reproduction, but also reduces the dimensions of the latter. Social reproduction refers to a wide variety of social practices such as house cleaning, domestic tasks, the socialisation of children and so on which are necessary for both the physical and psychological reproduction of skills and capitalist values and the maintenance of a consumer market (Mackenzie 1979). As these activities usually take place within the family they are viewed as 'natural' (and private) and have traditionally been attributed to women, allowing these social practices to be characterised as apolitical.

The effect of this conflation of various aspects of reproduction into collective consumption has been to marginalise large areas of human activity and experience. In urban politics it has led to gender being recognised only in so far as the production/consumption dichotomy is seen to echo a male/female divide: hence, Castells' (1978 p.178) remark, 'the contemporary city also rests on the subordination of "women consumers" to "male producers" '. Issues of consumption do have a particular salience for women, but gender roles cannot be mapped directly onto this dichotomy. To understand how gender relates to these categories we must return to a much fuller notion of reproduction and, most importantly, consider the dialectical relations of production and reproduction. These are not discrete processes or structures amenable to separate analysis, but constitute an essential unity, forming, as it were, two sides of the same coin. Mackenzie (1979 p.19) makes the point that this analytical separation of production and reproduction is

> ...the basis of Castells' subsumption of reproduction within the concept of collective 'consumption'. Shorn of its integral structural connection to production, reproduction becomes a theoretically isolated process, and only its most obvious surficial relations those of consumption remain. The complex relations and content of reproduction of labour become perceived as a process of individual and collective consumption, dominated by the

latter.

This serious misconstruction of existing analyses of urban politics stems from the failure to recognise that the family has its own political and economic reality within the domestic economy. The constant focus of attention on collective consumption processes has effectively squeezed the domestic economy and its central activity and process of reproduction out of the frame of analysis. Viewing the domestic economy as a self-sufficient sector, with no integral links to the broader market economy, has led to the emergence of certain conceptual errors such as the consideration of housewives as an industrial reserve army of labour that can automatically be drawn upon by the market economy (Garabaghi 1983). The family has its own structural requirements and constraints, and the labour 'employed' in the domestic economy can only be released if there is an internal restructuring of that economy, as happened in the post-war period in Britain. This historical restructuring of the relations between the domestic and market economy has been ignored in urban politics which has turned all its attention to its corollary i.e. the combined restructuring of public and private provision outside the domestic economy. Thus, the growth of consumption goods and services has been viewed from the perspective of their source of provision rather than being seen as a manifestation of the increasing integration between the domestic and market economies.

This process is revealed by certain trends which illustrate the manner in which women's reproduction role mediates their involvement in production (see Bondi and Peake, forthcoming). The total amount of time women may spend out of the labour force having children obviously varies according to the number of children they have and the length of time between births. A comparable factor for all these women, however, it the median time of return to work after the latest birth and this has gradually diminished over the post-war period (Martin and Roberts, 1984). The reduction in time spent on biological reproductive processes is the result of women having smaller families spread over shorter periods of time allowing reciprocal changes in the sphere of production, such as the increase in the proportion of women returning to part-time work. The unprecedented expansion of employment for women, particularly married women, since the 1950s occurred largely in the public services with jobs for clerical workers, domestics, nurses, teachers, social workers, and so on. Concomitantly, women's entry into waged work, resulting in less time spent in the home, led to an increasing demand for socialised reproductive services such as nurseries, school meals, laundries, fast food etc. It is women's wages which have funded the increase in the new consumer goods and services being provided. The growth in women's participation in paid work, together with its corollary, the growth of state intervention in social reproduction (commonly seen as collective consumption) was restructuring relations of production and exchange within households and between family members and the state. These trends began to change with the onset of the current recession. Whilst cuts in socialised reproduction provision affect everyone, they particularly influence the lives of women, both in their roles as producers (employees) and as the primary users of services. Cuts in social services, for example, involve a transfer of functions from the state to the family within which extra tasks almost invariably fall to women. They also involve a transfer from a collective (public) to an individualised (private) mode of consumption, which, as emphasis has been placed solely on the former, has not been seen as problematic. It is to these relations between the public and private spheres that we now turn.

Whilst the entry to a much fuller notion of reproduction into its frame of reference is a necessary development for urban politics, an emphasis on reproduction is insufficient, for another feminist imperative which requires addressing. We need to go beyond the treatment of the family solely as an economic unit viewed from the requirements of capital to consider both the complexity of social practices and relations within the family and the meaning which they impart to women(3). The conceptual framework most fruitful for this analysis is that of the public/private divide, which Garmanikov and Purvis (1983 p. 5) refer to as 'a metaphor for the social patterning of gender'. It is the false association of women with the private world of the family and interpersonal relations and men with the public world of politics and employment which allows urban politics to classify conventional female domestic activities as irrelevant for the transformation of social relations and outside the bounds of study. We need to explore and reconstruct our notions of the public and private in order to alter our conception of political activity.

The public/private division rather than serving as an accurate description of social organisation has been utilised as an ideological device which obscures discrimination against women (Elshtain, 1981). The feminist claim of the 'personal is political' has attempted to rectify this by recognising that women's activities at the inter-personal level, through the search for new, progressive social identities are a profound source of social change. However, this all encompassing motto does not serve to differentiate between certain activities and relationships which a political perspective demands. The conceptual categories, public and private, give structure to various situations and cannot be dismissed as mere arbitrary conventions that we need to dispense with forthwith. If there are no distinctions between public and private spheres of life then politics becomes a redundant word, devoid of meaning. Without this distinction we would all be viewed, and assessed, in our roles as individual citizens and would thereby devalue the social relations of the private world by depriving the latter of its meaning. Instead we need to challenge the isolation and discrimination of women within male dominated practices and social structures which provide for the oppression of women and their association with the private sphere. With respect to the private world we need to ask if there are aspects of familial ties which are so fundamental to human and social existence that we cannot reject them. Recognising that the private familial sphere has its own values and sense of purpose is to affirm that certain experiences and social relations within it have their own meaning. These meanings have evolved over time with changing social practices and relations, and will lead ultimately to an altered view of human life and reality. Thus, we should distinguish between those activities and relationships and experiences which are crucial to the formation of social identities, and those which can (and therefore should) be transformed. This requires recognising and teasing out the two elements contained within the family sphere: the family as a permanent kin network i.e. what it means to be a human being, and the family as the site of social reproduction and oppression of women i.e. what it means to belong to the gender category 'women'.

Studies of urban politics can break the mould not only by ascertaining the structural constraints determining women's participation in urban political activities but also by recognising that integral to this task is the meaning that the private world of the family plays in women's participation, for as Harstock (1979, quoted in Brownill) claims:

> A mode of life is not divisible. It does not consist of a public part and a private part, a part in the workplace and a part in the community - each of which makes up a certain fraction, and all of which add up to 100%. A mode of life and all aspects of that mode of life, take meaning from the totality of which they form a part.

The following section reveals how the lack of concern with the feminist imperatives addressed in this section has seriously undermined both our understanding of urban politics and of the gender category 'women'.

Some Observations on Methodology

Little research on the subject of women and urban politics has been carried out in this country, and even less in the developing world (see Moser n.d. for a developing world example). Yet women's increasing involvement in urban politics is evidenced in both the formal political arena - voting and political institutions - and in informal political activities. The purpose of this section is to examine critically the assumptions and biases that influence the treatment - and this has been partial - of women in studies of urban politics. Aspects of both the formal and informal arenas of urban politics are examined as it is considered important, as Dunleavy (1979) recognises, to breach the artificial separation of electoral analysis from wider aspects of urban politics in order to promote explanations of alignments based on a broad understanding of social processes(4). Lack of space, however, prevents me from reviewing the position of women in formal political institutions and restricts the analysis to voting and locality based politics.

Some feminist social scientists have criticised studies of voting behaviour on the grounds that information collected from social surveys is

70

inherently anti-women, and, therefore, should not be used to analyse issues concerning women (Graham 1983). Be that as it may, survey methodology is a necessary tool in any analysis involving large numbers of women which requires a representative sample in order to be able to make generalisations about the population under study. Quantitative analysis, it has been claimed, is alienating; however, in itself it is not oppressive, and the notion of female aversion to numbers as a precondition of feminist research has no logical basis (Bowlby, Peake and Wilkins 1983). Nevertheless, the quantitative era was associated with an empirical bias which impeded the development of feminist theoretical advances. As Lovenduski and Hills (1981) acknowledge, political science has concerned itself almost exclusively with the empirical study of public power, political elites, and the nature of government institutions, thereby excluding women who are largely invisible in these spheres. Concomitantly, given the empiricist bias of study it was not usual for this absence to be questioned. Voting behaviour studies have also been criticised for promoting the male bias of areas of study (Lovenduski and Hills, 1981). On the grounds that women tend to participate less than men in formal political activities, critics have claimed that attention should focus instead on the informal political activities of the private sphere. As mentioned earlier this view fails to recognise the distorting ideological function of the public/ private divide: women have a high level of involvement in the public world in waged work, and,

> There is also evidence that instrumental considerations may make traditional political issues more salient to women and voting behaviour studies are a good initial indicator of such a change. We would contend that women's support for political parties and the legitimacy governments receive from women's votes merits attention...
> (Lovenduski and Hills, 1981, p.2.).

Elections are also of obvious importance in determining the provision and distribution of various goods and services (Peake 1984). For these reasons the potential significance of 'consumption locations' on political alignments is considered to be an important area of study.

   To illustrate the inadequacy of an approach

which fails to recognise that the specificity of women's political position derives from the fact that they have work roles in both the spheres of production and reproduction (Mackenzie 1979) it may be useful at this stage to examine a particular study of voting as an urban political activity. Patrick Dunleavy's (1979) study of the urban basis of political alignment builds on the progressive breakdown of the relationship between class and party voting, which, he argues, is well documented but inadequately explained. He claims that this is due to the insistence on explaining the post-war decrease in class voting in terms of the declining salience of occupational class itself rather than suggesting that structural changes are creating a new cleavage which is cross-cutting that of occupational class and reducing the latter's impact on political alignment. Essentially his approach recognises that structural changes cannot be assimilated into occupational class explanations. In his analysis of Gallup surveys of February 1974 he investigates the influence of housing and transport locations on political alignment in a convincing attempt to demonstrate that a 'sectoral consumption approach provides the most comprehensive and compelling account of the role of urban cleavages in political alignment' (1970, p. 424). In the sphere of consumption Dunleavy argues that individuals interests are structured by their location in either the public or private sector. This sectoral divide does not follow the boundary between working- and middle-class, but cuts across and consequently fragments the social bases of political party support, in particular that of working-class support for the Labour Party.

In many ways Dunleavy's formulation is attractive, but it is also inadequate, allowing him to ignore the particular significance of individual and collective consumption for women's daily lives and the differential impact of 'consumption cleavages' on men's and women's political allegiences. Although he expresses dissatisfaction with the social grades used by Gallup he does not appear to view the use of the household as the basic unit of analysis as problematic. His investigation of private transport provision, for example, fails to appreciate that within the household there is differential access to cars. An increasing number of studies, however, are developing an appreciation of the impact of gender constraints on transport provision (Pickup 1982). A questionnaire survey of 190 respondents on three housing estates in Reading revealed that access to vehicles varied within households (Peake, 1983). Whilst 53.5 per cent of households had access to a car only 38.7 per cent of respondents used a car. Overwhelmingly, reasons given to explain the discrepancy between a car being available and a car being used revealed that it was the women in the sample who did not have access to a car, and consequently, relied on public transport to a much larger extent than their husbands. Furthermore, within different housing tenure categories these discrepancies are even more marked(5).

Whilst Dunleavy is absorbed in looking for variations in consumption locations across social class categories he fails to comprehend that there are also significant variations within consumption locations according to gender. Dunleavy is adopting a mode of enquiry which places certain women respondents into categories which do not relate to the social reality of their own daily lives, but to those of their husbands(6). Furthermore, in applying quantitative data in isolation from the attitudinal data obtained from the respondents he is ascribing meanings to results which may not appertain to the respondents in question. The categories used in his analysis are a reflection of assumptions and modes of thought of dominant groups, and therefore do not necessarily present the social reality of the daily lives of women and the meanings that their actions have for them (Martin and Roberts 1984).

It is not an easy task to interpret changes in women's voting patterns as the manner in which voting data are usually assembled is not conducive to treating women as individual citizens (Lovenduski 1981). In the Gallup survey used by Dunleavy married men and unmarried respondents were classified according to their own occupation whilst married women, even if in paid employment, were classified according to their husband's occupation(7). However, if the basis of a system of stratification is occupation then an individual cannot also be stratified according to marital status; it is inconsistent to take paid work into account for single women but not for married women. Moreover, it is illogical that the social position of women who do not have paid jobs is not taken into account as a factor of their own class position when their social position is considered both a sufficient and necessary reason to attribute them to the class of their male kin (Delphy 1984). The impact of the feminisation of the work force, the transformation of the occupational structure and changing household structures, and the significance of these links for class structure and voting patterns is totally disregarded as a result of this method of categorising female respondents. For example, women in non-manual occupations, married to men in manual occupations are classified as working class respondents, yet as men in the same jobs they would have been classified as middle class. Hence, many of Dunleavy's working class respondents whom he claims have consumption locations which contradict their class positions may not be males in manual occupations but women in non-manual occupations whose consumption locations consolidate their class position.

Placing women into the class stratification system by virtue of their marital status results from the dominant mode of thought which views the family as the 'natural' unit of classification. Hence, comparisons within the family unit have been considered unnecessary, and are indeed impossible; women are automatically attributed the status of their male kin. It follows from this that wives are assumed to be of equal status to their husbands, hence the question of women's dependence on men is obscured. That is, the fact that women are not equal to men in many ways is not seen as a relevant criterion for stratification purposes (Acker 1973). Male dominance and 'masculinity as the norm' have led to the further assumption that women's voting preferences are merely extensions of their male kin's. These assumptions cannot be attributed merely to technical or methodological errors; they reveal an epistemology in which women are regarded as being of inferior status(8). Whilst it is necessary to pose solutions to these problems at the level of method - such as ascribing every woman a place in the class stratification system regardless of marital status - it is also essential to consider broader social and economic relations which ascribe women a primary role of reproducer, and to reconceptualise the gender category 'women' in order to accord women's interests and needs a status equal to that of men's.

The emphasis on studying women's participation in formal spheres of activity, and its often empiricist nature, has served to divert attention from the fact that women's perceptions of themselves are rooted in family life and that women are becoming increasingly participation oriented. As socialised reproductive services have become more central to domestic and community life, the content and conditions of women's domestic labour have become more of a political issue. Consequently, women have become increasingly involved in struggles around issues of social reproduction. Whilst there are obvious variations in the type and amount of activities engaged in between women there has been, overall, an increasing demand on the part of women to influence the public policy decisions that structure the daily content of their lives (see Lovenduski and Hills 1981). A large number of women have taken part in urban political activities and belong to a wide variety of social organisations, some of which act through the political system and others which are concerned with issues often labelled as 'apolitical' e.g. demanding zebra crossings, play groups, communal play space for children, train and bus services etc. (see Peake 1983).

Why is it that women are most active in this sphere and yet have remained outside the attention of urban politics? What happens to our understanding of urban politics if we assume women are as interested in and as competent to exercise political power as men? Owing to androcentricity being upheld as the norm for political behaviour, the fact that the political experience of women is not the same as that of men has often resulted in the view that women are less competent than men (Bourque and Grassholt, 1974). This has not only led to the interpretation of women's participation in the informal sphere of urban politics as misplaced, unimportant and irrelevant, but also to the political implications of women's distinctive experience being overlooked (Ettore 1978). Women, it seems, may be perceived as part of the social structure, but not as part of the structure of power (Rendel 1981) as power is seen to reside within the state arena where decision-making over the provision of collective goods and services takes place.

The factors which result in the separation of the politics of collective consumption from mainstream class politics have been seen by some to limit the potency of popular protest action around these issues (Bondi and Peake, forthcoming). Saunders, for example, argues that 'those whose primary concerns lie in furthering the conditions for the development of socialism will derive little return from either analysis of or activity in urban politics' (1981, p. 276). This mode of enquiry effectively disregards both the political meanings attached to women's involvements and the study of how and why women come to have power (albeit a small amount) within the community. Women's involvement in informal community activities has not been of general interest unless they become institutionalised as pressure groups (or even better as urban social movements). A basis does exist, however, for taking a more positive view of gender-based activities in urban politics. Firstly, any attempt to analyse the potency of gender-based struggles within existing categories, such as 'urban-social movements', which retain an inadequate conceptualisation of reproduction, will, by their very nature, be partial and incomplete(9). From a feminist perspective, McDowell remains sceptical of the radical potential of community-based political action, claiming that

> The continuing strength of the ideology of the home as haven rather than as workplace, its significance as an object of conspicuous consumption and as a status indicator, the local provision of collective goods and services all reduce the prospect of gender-based urban social movements (1983, p. 69).

An emphasis on consumption leads to the failure to recognise the urban arena as the site of social reproduction and, thus, sees women's participation in urban political activities as that of consumers, rather than as reproducers i.e. as workers.

Secondly, whilst data are difficult to compile it is obvious that women are extremely well represented in a wide array of informal political groups: local protest groups, mass campaigns, community action groups, welfare campaign groups and so on. In addition, women participate in enormous numbers in all-female organisations: in Britain over three million women belong to such organisations as the Women's Institute, the Townswomen's Guild, the Women's Royal Voluntary Service and the Mothers Union (Randall, 1982). These women's associations form an important community-based network for women's influence across a wide range of issues - both public and private. Their existence demonstrates both the organisational reality of community life and the salience of gender as a basis for organisation. The political significance of these networks, however, is only made explicit when reproduction in explicitly treated as a work process: the community as workplace is the focus of political activity arising out of issues of reproduction in a manner analagous to that of the shopfloor in connection with production issues (Bondi and Peake, forthcoming). Moreover, these groups provide an opportunity for a much greater degree of role continuity between 'political' and 'non-political' activities, which is particularly important for women for whom domestic roles have great salience. The rooting of women's consciousness in the family allows for the potential conflict between their caring, maternal roles and political activism around issues of reproduction to be dissipated. These groups cannot be dismissed out of hand before assessing the meaning they have for their participants(10).

Finally, by challenging the conception of politics as a public activity, largely carried out in the state arena by men, these forms of political action challenge conventional definitions of politics (and ultimately of the socialist cause) and alter the site of the political arena. The growth of community-based struggles, together with the contemporary peace movement (most notably Greenham Common and Mothers for Peace) provide evidence of the way in which gender influences the development of political consciousness - and in particular the divide between the public and the private areas of life - and the form of political mobilisation. More traditional political movements have been slow to recognise the transformative potential of these newer groups, but there are signs of change (see, for example, the Labour's Lost Millions Debate in New Socialist, together with the political direction of the GLC). As Mackenzie (1979, p. 17) asserts

> Their [women's] position at the centre of the dialectic between production and social relations of reproducing labour power may make women pivotal in formulating a totalistic understanding of social change as such more than a redistribution of wealth or a change in the ownership of the means of production, but change which transforms the entire fabric of social relations.

Conclusion

The underlying imperative of the discussion on the androcentric biases and false assumptions which have permeated empirical studies of women and urban political activities was to argue for a transformation of the methodological and conceptual bases of urban politics. A mode of inquiry that enables us to stop treating women as objects, and see them as the subject of inquiry i.e. as participants, not respondents would enable us to work towards this goal. This involves adopting a method which aims to help women become conscious of their social positions through a process of political discourse which they actively engage in and which helps to create new meanings of what it is to be a woman. Political discourse can be seen as a search for new identities and for truth - truth defined as the active creation of meaning between two or more people through the contestation of old ways of understanding and meaning. If we aim to transform social identities and relations in such a way that gender categories are no longer oppressive then a primary objective must be to allow for this process of critical reflection among participants and between participants and researchers. Not doing so leads to a tendency to treat women as abstract persons, taking their social positions for granted and, ultimately, losing touch with the values and imperatives produced from their private worlds, and hence, with everyday concrete meaning. Critical examination of social relations and practices at the point of reproduction involves examining the nature of the work performed by women (and men and children) and the relations of production and exchange that are generated by this work (see Elshtain 1981). This mode of inquiry should enable us to begin reconceptualising the gender categories of 'women' and 'men' and the activities of 'production' and 'reproduction' which

> ...in turn presupposes and reaffirms a new understanding of social change. Changes in our activity patterns and in our relationships with other people are not a by-product of social change and of the struggle to create and control this change. They are the very stuff of which struggle and change are made, and toward which these move (Mackenzie, 1980, p.21)(11).

NOTES

1.   Women's political experience has been
devalued in four ways: the ommission of women as
subject matter; women being recognised in relation
to men rather than in their own right; the assumption of male behaviour and masculinity as the norms
of political practice; and, the exaggeration of
gender differences.
2.   The discussion in this paper is concerned
solely with the development of urban politics in
Britain, and as such the comments made should not
be taken as pertaining to other Western societies
where there are significant differences in their
conceptions of public and private spheres of life.
3.   At this point it is necessary to say
a few words about the manner in which this paper
addresses the question of meaning. In making an
analytical distinction between labour   (system
integration) and social interaction (social integration), Habermas (1976) refuses to reduce practice
to activities contained within categories of labour.
Practice is also viewed as consisting of a 'form
of interaction between people which presupposes
some kind of understanding (not necessarily correct)
of social meaning...' (Sayer, 1983, pp. 14-15).
Sayer further claims that

> Meaning is normally thought of as   purely
> descriptive, as a label for something external
> to us.   But it is also 'constitutive' of
> human practice insofar as what we do depends
> on understanding meanings or concepts which
> are available in society... Take the practices
> associated with masculinity and femininity
> and their   constitutive   concepts. 'Feminine
> behaviour' both confirms and reinforces, and
> is informed by everyday concepts of what it
> is to be feminine; they are 'reciprocally
> confirming' as Williams puts it.   This also
> applies to questions of identity, of what
> it is to be a husband, housewife, Palestinian
> or whatever.   Our identity is shaped by what
> we can do, and what we can do is shaped by
> how others act towards us, on the basis of
> their assumptions about our identity and about
> their own (pp. 13-14).

Within the study of urban politics we need to recognise the important role of transforming individual's
positions, not only in the labour process, but

also through their interaction with other subjects i.e. through their own social identities. Only by addressing both of these aspects of practice will we be able to work towards genuine political emancipation.

4. Whilst electoral data may give only a narrow and incomplete picture of political alignment we cannot afford to ignore it on these grounds alone (Dunleavy 1979). Moreover, as Castells (1978, p. 7) asserts, elections 'lie at the heart of the liberal democratic state'.

5. These discrepancies are revealed in table 5.

6. Similarly, in following articles which examine the relationship between production locations of union/non-union and public/private sector employees and social class, Dunleavy (1980) ascribes trade-union membership to respondents if any member of their family was a member of a union.

7. Delphy (1984) points out that in certain French studies even this derogatory treatment has not been consistently applied within the same study.

8. These practices employed by Dunleavy and others such as Särlvik and Crewe (1983) have resulted over the years in various fallacious assumptions concerning women and voting which have largely been discredited (although they still tend to have a wide currency). Some of the more common ones are outlined below. In Britain prior to 1979 it was assumed that women were more likely than men to vote Conservative. This voting pattern was later attributed to a generational factor i.e. a higher proportion of women amongst the elderly, revealing differential voting patterns amongst women, and also discrediting the view that voters became more conservative as they grow older. Another conventional wisdom put forward was that women vote less than men. Recent feminist research, involving a reworking of Butler and Stokes' (1974) data has shown differences to be minimal. In 1964, 90% of eligible women and 92.5% of eligible men voted, whilst in 1970 the corresponding figures were 83.3% and 87.1% (Baxter and Lansing 1980). At least since the 1960s British women have voted at the same rate as men (Hills, 1981). Further, Himmelweit et al (1981) have repudiated the thesis that women voters will base their voting decisions less on evaluations of the policies at stake, and more on their husbands' evaluation. In a re-analysis of Butler and Stokes' (1974) 1970 sample and Crewe's (1974) sample they found that issues were

Table 5:   Households With and Without Access to a Car
            on Three Housing Estates in Reading

	Total Sample	Owner Occupied Estate	Council Estate 1	Council Estate 2
Household with Access to a Car	53.5%	86.0%	48.8%	25.8%
Respondents with Access to a Car	38.7%	72.0%	30.4%	13.8%

Source: Peake (1983)

equally relevant to women and men voters and that there were no differences in the issues that mattered. This is not to claim that women do not differ at all from men in the way they vote. Francis and Payne (1977), also in a re-analysis of Butler and Stokes' 1964 and 1970 general election data, found that women in all age cohorts departed from their model's expectation more than men.

9. This criticism can also be applied to the study by Ettore (1978) of the women's liberation movement as an urban social movement.

10. Pickvance (1975) claims that this requires an examination of the structure of social relations in the 'social base'. This involves looking not only at organised groups and institutions, but also at the informal ties within the community of kinship, friendship etc. In addition there is a need to study the value orientations relating to women's motivations for participation. Study of other influences on participation, such as short-age of time, will help to bring women's role as workers, both in production and reproduction, to the fore.

11. Activity patterns are taken here to refer to both the transformations of material things and the sharing of meaning (see Sayer 1983).

REFERENCES

Acker, J. (1973) 'Women and Social Stratification: a case of intellectual sexism', American Journal of Sociology, 78, 936-45.

Barrett, M. (1980) Women's Oppression Today, Verso, London.

Baxter, S. and Lansing, M. (1980) Women and Politics: The Invisible Majority, University of Michigan.

Bonder, G. (1983) 'The study of politics from the standpoint of women', International Social Science Journal, 98, 3, 569-84.

Bondi, E. and Peake, L. (forthcoming) 'Gender and Urban Politics' (Paper to be included in forthcoming Women and Geography Study Book.)

Bourque, A. and Grassholtz, J. (1974) 'Politics as unnatural practice: political science looks at female participation', Politics and Society, 4, 2, 225-66.

Bowlby, S., Peake, L. and Wilkins, P. (1983) 'Doing feminist research in geography: a discussion' Meeting of Minds? Relationships Between Feminism and other Modes of Thought in

Geography, Women and Geography Study Group, Institute of British Geographers, pp. 18-23.

Brownill, S. (1981) 'A Women's Place Is? An Exploration into the Relationship Between Women and the Community', mimeo, C.U.R.S., Birmingham University.

Brownill, S. (n.d.) 'From Critique to Intervention: Socialist - Feminist Perspectives and Urbanisation', mimeo, C.U.R.S., Birmingham University.

Butler, D. and Stokes, D. (1974) Political Change in Great Britain. Forces Shaping Electoral Choice, Macmillan, London.

Castells, M. (1978) City, Class and Power, Macmillan, London.

Crewe, I. (1974) 'Do Butler and Stokes really explain political change in Britain? European Journal of Political Research, 2, 47-92.

Delphy, C. (1984) 'Women and Stratification Studies' in C. Delphy, Close to Home. A materialist analysis of women's oppression, Hutchinson, London, pp. 28-39.

Dunleavy, P. (1979) 'The urban basis of political alignment: social class, domestic property ownership and State intervention in consumption processes', British Journal of Political Science, 9, 409-43.

Dunleavy, P. (1980) Urban Political Analysis, Macmillan, London.

Elshtain, J.B. (1981) Public Man, Private Women. Women in Social and Political Thought, Martin Robertson, Oxford.

Ettore, E. (1978) 'Women, urban social movements and the lesbian ghetto', International Journal of Urban and Regional Research, 2, 499-520.

Foord, J. (1983) 'Introduction in Meeting of Minds? Relationships Between Feminism and Other Modes of Thought in Geography, Women and Geography Study Group, Institute of British Geographers, pp. 1-4.

Francis, J. and Payne, C. (1977) 'The use of the logistic model in political science: British elections 1964-1970', Political Methodology, 4, 233-70.

Garabaghi, N. (1983) 'A new approach to women's participation in the economy', International Social Science Journal, 98, 659-82.

Garmanikow, E., Purvis, J. (1983) 'Introduction' in E. Garmanikow, D. Morgan, J. Purvis and D. Taylorson (eds.), The Public and the

Private: Social Patterns of Gender Relationships, Heinemann, London, pp. 1-6.

Graham, H. (1983) 'Do her answers fit his questions? women and the survey method' in E. Garminakow, et al (eds.), The Public and the Private: Social Patterns of Gender Relationships, Heinemann, London, pp. 132-45.

Habermas, J. (1976) Legitimation Crisis, Heinemann, London.

Harstock, N. (1979) in Z. Eisenstein (ed.), Capitalist Patriarchy and the Case for Socialist-Feminism, Monthly Review Press, New York, pp.

Hills, J. (1981) 'Britain' in J. Lovenduski and J. Hills (Eds.), The Politics of the Second Electorate, Routledge and Kegan Paul, London, pp. 8-31.

Himmelweit, H., Humphreys, P., Jaeger, M. and Katz, M. (1981) How Voters Decide. A longitudinal study of political attitudes and voting over fifteen years, Academic Press, New York.

Hindess, B. and Hirst, P. (1977) Mode of Production and Social Formation, Macmillan, London.

Lebas, E. (1981) 'Introduction' in M. Harloe and E. Lebas (eds.), City, Class and Capital, Arnold, London, pp. ix-xxxiii.

Lovenduski, J. (1981) 'Towards the emasculation of political science: the impact of feminism' in D. Spender (ed.), Men's Studies Modified, Pergamon Press, Oxford, pp. 83-97.

Lovenduski, J. and Hills, J. (eds.) (1981) The Politics of the Second Electorate, Routledge and Kegan Paul, London.

Mackenzie, S. (1979) untitled mimeo, University of Sussex.

Mackenzie, S. (1980) 'From the "New Citizenship" to "Beyond the Fragments" ', mimeo, University of Sussex.

Martin, J. and Roberts, C. (1984) Women and Employment. A Lifetime Perspective, Department of Employment and Office of Population Censuses and Surveys, HMSO.

McDowell, L. (1983) 'Towards an understanding of the gender division of urban space', Environment and Planning D: Space and Society, 1, 59-72.

Moser, C. (n.d.) 'Residential Level Struggle and Consciousness: The Experiences of Poor Women in Guayaquil, Ecuador', mimeo, Development Planning Unit, Bartlett School of Architecture and Planning.

Peake, L. (1983) 'The Role of Consumption Processes in Electoral Geography and their Impact on Political Cleavages', unpublished Ph.D. thesis, University of Reading.

Peake, L. (1984) 'Why Särlvik and Crewe fail to explain the 1979 election result and electoral trends in the 1970s', Political Geography Quarterly, 3, 161-67.

Pickup, L. (1982) 'Combining family and work roles - a time-geographic perspective on women's job chances', Paper presented at the I.B.G. 1982 Conference at the Women and Geography Study Group Session on The Institutionalisation of Gender Differences.

Pickvance, C. (1975) 'From "Social Base" to "Social Force": Some Analytical Issues in the Study of Urban Protest', C.E.S. CP14 Proceedings of the Conference on Urban Change and Conflict, University of York.

Randall, V. (ed.) (1982) Women and Politics, Macmillan, London.

Rendel, M. (ed.) (1981) Women, Power and Political Systems, Croom Helm, London.

Särlvik, B. and Crewe, I. (1983) Decade of De-alignment. The Conservative Victory of 1979 and Electoral Trends in the 1970's. Cambridge University Press, London.

Saunders, P. (1981) Social Theory and the Urban Question, Hutchinson, London.

Sayer, A. (1983) 'What kind of marxism for what kind of feminism?' in Meeting of Minds? Relationship Between Feminism and Other Modes of Thought in Geography, Women and Geography Study Group, Institute of British Geographers, pp. 13-17.

CHAPTER FIVE

LITTLE GAMES AND BIG STORIES: ACCOUNTING FOR THE
PRACTICE OF PERSONALITY AND POLITICS IN THE 1945
GENERAL ELECTION

NIGEL THRIFT

> Children, only animals live entirely in the
> Here and Now. Only nature knows neither mean-
> ing nor history. But man - let me offer you
> a definition - is the story-telling animal.
> Wherever he goes he wants to leave behind
> him not a chaotic wake, and an empty space
> but the comforting marker-buoys and trail-
> signs of stories. He has to go on telling
> stories. As long as there's a story, it's
> all right. Even in his last moments, it's
> said, in the split second of a fatal fall
> - or when he's about to drown - he sees, pass-
> ing quickly before him, the story of his whole
> life (Swift 1983, pp. 53-54).

Introduction

This chapter represents the third of a series of
papers inquiring into the nature of social action.
The first paper (Thrift 1983b) was concerned with
providing a general theory of social action, con-
ceived as a situated and never-ending discourse
driven by conflict. The second paper (Thrift 1985a)
took up one necessary component of any theory of
social action, the geography of knowing and un-
knowing within which social action must be developed. This paper is concerned with another
such component, the human agent. For as Giddens
(1982, p.535) puts it, 'formulating a theory of
social action in the social sciences demands
theorising the human agent'. 'Understanding the
agent' is not a modest aim. Clearly, in a short
paper like this one, I cannot and will not present
any finished or even half-finished theory of the
human agent (see Thrift 1987). Like the other
two papers, this paper is 'pre-theoretical'.
    Still there are important issues to be
addressed. On the theoretical front there is the
problem of how it is possible to interpret a whole
constellation of terms which are routinely used
in social science but whose routine usage conceals
very considerable problems - terms like

'experience', 'consciousness', 'ideology', 'action', 'belief', 'self', 'attitude', and so on. On the practical front, there is a problem of what is being talked about when political consciousness (and beliefs and attitudes) is considered, and under what conditions political consciousness can be changed(1).

The paper is in two main parts. The first part of the paper sketches an outline of how human agents might legitimately be theorised through a discursive model of how people build themselves through others and what it is that they build. The second part of the paper narrows the terms of the discussion a little by concentrating on the problem of how 'political'(2) beliefs are built up and changed. I take as an example the case of the shift in political attitudes towards the left in England in the Second World War.

Telling It Like It Is: A Discursive Model of Human Agents

What is a human agent? This is a short enough question but it requires a long answer.· What I want to suggest is that in many contemporary theories of social action, the conception of the human agent is deficient.

Human Agent/Negative Ascriptions

Theories of social action must include within them a more or less explicit conception of the human agent. In many current theories of social action, the conception is deficient, usually because the human agent appears in the theory by default, very much as the ill-considered trifle (Geras 1983). Three deficiencies are particularly common.

The first of these deficiencies is one that is now less common than it was, but it is by no means laid to rest. This is that the human agent can be reduced to a cognitive drone, to a string of internal programmes responding to an external environment. People's action is governed by some 'inner', on-board computer which assimilates all the available knowledge, works out the angles and decides on an appropriate course of action. It has surely been argued often enough by now that human beings are not rational beings, at least in this sense of rationality. People live in a social world and they cannot therefore be reduced to episodes of theoretical reasoning, followed by appropriate action. This is an incorrigibly contemplative view of the world which flies in the face of the evidence; evidence which suggests that people act to reason to act. Such a reduction is only possible because the 'intellectual fallacy' is still so pervasive in social science; a fallacy which enables middle-class academics to pass off 'their curious customs and those of their students as the psychology of all mankind' (Harré 1983, page 260).

The second deficiency found in certain theories of social action is that the human agent can be reduced to a moral incompetent. Human agents are partly knowledgeable, creative (rule making as well as rule following) and responsible beings who - under pressure from their peers - make judgements, evaluate and give accounts of themselves. To say that there are limits on the kind of judgements, evaluations and accounts people can make is not to absolve them of the responsibility for their actions (see Parfit 1984).

The third deficiency, found especially but not exclusively in marxist and neo-marxist theories of social action, is that the human agent can be reduced to the outcome of whatever is the current positive act. Conventionally this is achieved through a category like 'ideology'. In the most rigorous case there is the Althusserian notion that ideology interpellates individuals as subjects. Less rigorously, the category of 'ideology' can be used simply to indicate that a certain set of beliefs or preoccupations can be linked to the material situation of a particular social group. Both cases generate problems. In the most rigorous case the major problem is that no room is left for human beings as creative agents. Ideology functions, 'as the secret police of the social structure, arresting the suspects and shoving them into the correct cells' (Connell 1983, p. 227). In the least rigorous case, the problem is that there is no guarantee that particular beliefs and preoccupations are tied to particular material situations, so that the need for the category disappears. Further, the whole question of what 'beliefs' and 'preoccupations' actually are is left unanswered. Thus even used in the least rigorous way the category 'ideology' 'proves inert and unilluminatingly reductive' (Stedman Jones, 1983a, p.18). In the case of a set of political beliefs and preoccupations like Chartism, for example, a preoccupation with ideology simply missed what was most urgent to explain about Chartism - its political character, the specific reasons for its rise and fall, its focus upon representation and its lack of interest in the demarcation of socio-economic status within the unrepresented. The difficulty of an explanation in terms of the limitations of an artisanal consciousness or ideology...was that it did not identify with any precision what it was that declined (Stedman Jones, 1983a, p.19).

One alternative to ideology has been the category 'hegemony' but it is difficult to see how, in many of its usages, a category like 'hegemony' escapes the problems posed by a category like 'ideology'. It 'may register some moral distance from the apologetic complacency of functionalist theory' (Stedman Jones 1983a, p. 86) but it makes no real break from the whole theoretical bloodline.

One answer to these problems has been to redefine ideology and hegemony in such a way that these categories become broad enough to escape the charges of reductionism. Thus Hall (1983, p.59) defines ideology as

mental frameworks - the languages, the concepts, categories, imagery of thought, and the systems of representation - which different classes and social groups deploy in order to make sense of, define, figure and render intelligible the way society works.

Similarly, Williams' (1977, pp. 112-113) defines hegemony as

a realised complex of experiences, relationships and activities, with specific and changing pressures and limits. In practice, that is, hegemony can never be singular. Its internal structures are highly complex as can readily be seen in concrete analysis. Moreover...it does not just passively exist as a form of dominance. It has continually to be renewed, recreated, defended and modified. It is also continually resisted, limited, altered, challenged by pressures not all its own.

The problem with definitions like these, of

course, is that in order to avoid a charge of reductionism they are so broad that few would disagree
with them. Further each definition is made up
of the undefined. All the work is left to
do(3).

These, then, are three of the most common
deficiences to be found in conceptions of the human
agent. If nothing else, they show that rationalism
still holds much of social science in a terrible
stranglehold; human agents are rational beings,
living in a theoretical world. But what of the
real social world, as opposed to this academic
world? What beings dwell there?

Human Agent/Positive Prescriptions

Luckily, there is a developing consensus in much
of philosophy and social science (in so far as
they can be separated) concerning what a human
agent is; a conception which theories of social
action must begin to take into account.

At the epistemological level, this consensus
invokes a rejection of the old concepts of rationality; 'science' is a problem not a solution. In
particular, there is a desire to recover a sense
of practical rationality (summarised in Berstein
1983, p.270). Old scientific modes of enquiry are
faulty:

> We must appreciate the extent to which our
> sense of community is threatened not only
> by material conditions but by the faulty
> epistemological doctrines that fill our heads.
> The moral task of the philosopher or the cul-
> tural critic is to defend the openness of
> human conversation against all those temp-
> tations and real threats that seek closure
> (Bernstein 1982, pp. 204-205).

At the ontological level, the consensus involves a new account of what a human being
is (though one with a very long bloodline)(4)
which is, essentially, the discovery that
hermeneutics is what people are and that psychology
is not a science but a humanity.

The depiction of the human agent built in
this new consensus is founded on a number of prescriptions. Three of these prescriptions are
particularly important. The first prescription
is that any depiction of a human agent must be
contextual, must accept that agents live pockets

of space and time and are not universals. There are a number of consequences of this prescription.

Human agents live a context which is predicated upon <u>action in time</u>, not contemplation. This is a fact of life; 'no psychological explanation of why a human being does one thing after another is required. We assume that human psychology is such that a person is continuously busy. It is a matter of energy exchange and biochemistry to explain why a human being is always in action (Harré 1979 p. 246). Upon action everything else is predicated.

1. 'Sometimes I don't know anything at all for large spaces: sometimes I know many things all in the same place. My perceptions are uneven, my understanding patchy but I have action; I go' (Hoban 1983, p. 38).

2. Human agents are developing systems that grow from simple individuals into richly structured ones. They are never complete. They are always growing. Human agents are 'dynamic structures' (Prigogine 1980; Prigogine and Stengers 1984) producing and reproducing themselves and others through a continuous flow of conduct and varying in how they are made up according to the vagaries of history, geography and the prevailing systems of social stratification that shape the contexts they live.

3. Human agents must live contexts that at any one time can only be partially determined, for in acting we do something; we make something take on a form other than that which it would have had if we had not acted; thus we determine the world. For this to be possible the world must be capable of being given a form which it did not already possess, that is, the world must be essentially indeterminate (Shotter 1984, p. 45).

4. Contexts are <u>active</u> networks of people gathered in particular social situations, not passive 'environments'. Action - in - context is therefore always <u>joint action</u>.

The second prescription is that language must be taken seriously as an <u>operator</u>, and not just as a parameter, as an <u>action</u> and not just as a representation - language is not just a passive, static framework through which 'experience' finds expression. It is a complex, context-dependent and therefore continually shifting rhetoric, that is a way of telling others what is being done in that context, 'binding together, in a systematic way, shared premises, analytical routines, strategic options and programmatic demands' (Stedman Jones 1983, p. 107). It is not a logic, but a way of making oneself understood to others:

> remember that, in general, we don't use language according to strict rules - it hasn't been taught us by means of strict rules, either. We are unable to circumscribe the concepts we use clearly; not because we don't know their real definitions, but because there is no real 'definition' to them. To suppose there must be would be like supposing that whenever children play with a ball they play a game according to strict rules (Wittgenstein 1965, p.25).

In this conception of language as a way of co-ordinating action:

> linguistic activities can be seen as working materially to structure social relations by 'in-forming', or instructing the participants in them in various ontological skills, in how to be parameters of this or that kind of activity, so that as persons they come to see and hear and act and do appropriate things in the appropriate contexts, 'routinely', 'naturally', one might say, as if without a moment's thought (Shotter 1985a, p.11, author's emphasis).

The third and final prescription is that human agents must be seen as socially constructing, not socially constructed. There is a world of difference between the 'ing' and the 'ed'. People are not just passively socialised into various social institutions. They are continually constructing these institutions and themselves and others anew - the three are difficult to separate - according to the particular context. Some examples of how social relations are continually constructed are in order. Let us start, appropriately enough with childhood, with the process of constructing of 'how to be' being encountered by a child for the first few times.

LITTLE GAMES AND BIG STORIES:

A mother looks at her young baby:

'Oh look', she says, after having got her
infant to look at her by cooing and smiling
at her, having placed her face in her line
of regard, "she's looking at me". So she
replies to her (?) look with a "Hello, hello,
you cheeky thing". The point here being that
whatever mothers do to motivate their babies'
activity, when they respond, mothers still
interpret their response as something which
their babies themselves do, not merely as
something which they have succeeded in eliciting from them; it is thus treated as activity
worthy of being an expression in a dialogue,
an expression requiring a meaningful reply...
in this situation babies can learn what they
bring about by their own actions (Shotter
1984, pp. 82-83, my emphasis).

Another example, a mother showing her eleven
month old child, Samantha, how to place shaped
pieces on a form-board.

"Having just physically helped her little
girl to place one of the pieces, Samantha's
mother says 'Oh clever girl!' But Samantha
had not paused in her activity and signalled
by eye-contact that she knew she had done
something socially significant; she just went
straight on to manipulating something else.
So her mother leant forward, caught her eye,
and repeated her 'marker' with emphasis:
'AREN'T YOU CLEVER?' Samantha then stopped
and smiled.
     Mothers are not satisfied with their
children doing the tasks that they require
of them. They must also give indications
in their actions that they did what they did
as a result of trying to do it, that 'they'
knew what was required of them, that their
actions were based in some knowledge of the
socially defined requirements of the situation.
Thus children must come to show in their
actions, not just awareness of their physical
circumstances, but self-awareness; an awareness
of their relations to others (Shotter 1984,
pp. 86-87, my emphasis goal).

Later on in life the mode of constructing
particular negotiated contents can become a powerful

resource.   A particular social group's 'how to
be' can be more confident, more manipulative. Take
the example of a tightly-knit group of the most
powerful Philadelphia families.   These are 'old'
families bonded together by interlocking director-
ships and their membership   of   the   exclusive
Philadelphia Club.

> These are families with a keen sense that
> their primary resources derive from their
> own past: the powerful mixture of historical
> distinction...and accumulated wealth. Because
> the very kernels of the families' conception
> of their social environment lies within them-
> selves, not the outer world, we could call
> these families 'solipsistic'...But these are
> also powerful, accomplished families who under-
> stand the underlying business structures of
> their city and know how to control those
> structures for their own ends...In an important
> sense, the families' ties to corporate
> directorships and their ties to the exclusive
> clubs provide complementary support to the
> families conceptions of themselves in the,
> world. The corporate directorships reinforce
> the families' sense of power and control;
> the social clubs reinforce the importance
> of their historical roots (Reiss 1981, pp.
> 299-302)(5).

The conception of the human agent that springs
from these prescriptions is very different from
that found in most theories of social action. Human
agents are not rounded theoretical beings, receiving
information, contemplating it, translating it.
This kind of depiction has only come about because
those who have studied social action have increasingly transferred features of their idea of discourse on social action to social action. Rather,
human agents are contextual beings, negotiating
each given context in joint action with other agents
with the aid of a particular store of practical
knowledge of whose meaning they are often unaware.
Thus the isolated individual ceases to exist:

> In other words, in many actions in daily life,
> we are ignorant as to what it is exactly we
> are doing, not because the 'ideas' or whatever
> is supposedly informing our conduct are too
> deeply buried in us somewhere to be brought
> out into the light of day, but because the

94

formative influences shaping our conduct are not wholly there in our individual heads to be brought out. Our actions occur interlaced in with those of others, and their actions are just as much a formative influence determining what we do as anything within ourselves (Shotter 1985a, p.15).

Storytime

Clearly the conception of the human agent outlined above has connotations for what we can regard as 'consciousness', 'belief', 'self', 'attitude', and all the other terms that are routinely applied to human agents, yet alone for how these terms can be stitched together into elements of a theory of social action like 'class consciousness', or 'ideology', or 'hegemony'. Just as clearly, as yet no strong theory of the human agent has come out of this conception of the human agent. Luckily, there is a model that is becoming generally accepted in disciplines as diverse as anthropology, sociology and social psychology, which begins the task of building such a theory, the so-called 'discursive' or 'constructionist' or 'constrictivist' model(6) with its five main elements - 'person', 'self', 'account', 'folk model' and 'intention'. The model is based upon the existence of human agents as contextual beings, structured out of joint action in joint action, and irredeemably social. Out of joint action human agents produce institutions which provide 'stocks of knowledge' (of various kinds but most especially the 'practical', 'tacit' or intuitive knowledge(7) that enables them to 'understand' what it is they are doing - saying - thinking). And these stocks of knowledge define their commonsense view of the world. Since at any point in time institutions will be structured in various ways, by social group and in space, so the knowledge institutions can impart will be differentially available to make persons.

So far, so good. But human beings (or rather, 'becomings') do not simply draw upon these stocks of knowledge, as if they were books on a shelf; for stocks of knowledge are not neutral; they come invested with the meaning of the interactions that obtained them and these interactions also learn the human agent how to be a person at the same time that the knowledge is imparted. (Indeed that is what much of the knowledge is about). So one learns to be a person over a long period of time in a process of progressive <u>self-specification</u>, the lessons of which are continually re-learnt, even modified, in interaction with other people in specific contexts. The person emerges; she is not there to begin with. Take the three examples cited above. In the first example the mother imparts her notion of 'baby' to the baby while in the second example the mother imparts her notion of being a competent girl to Samantha. In the last example the Philadelphia family creates persons competent to exercise power. And in each case, the activities that con-stitute these persons are going on out in the open, between people, rather than inside their heads, hidden from being directly per-ceived by others (Shotter 1984, p.23)(8). This is <u>social</u> being.

Two things follow from this depiction of person-making. First, it becomes clear that persons are differentially constituted within society. The opportunities for development of persons are differentially distributed through society - at home, at school, at work - so some human agents can grow into particular kinds of persons with particular kinds of resources while others cannot. There is a political economy of 'development opportunities'(9). Second, where the boundaries of the 'self', the 'inner core' of a person are drawn will vary too. There is persuasive evidence that ideas of self vary from society to society(10). There seems little reason, then, why there should not be a varying conception of self within societies. Thus, there is also a 'political economy of selfhood'.

How, given this radically social depiction of human agents and how it is that they grow into persons, is it possible to charac-terise the inner workings of individual human agents? In part, it is not. As the examples of the socialisation of young human agents as 'babies' and 'children' show what we take to be 'ourselves' is to a great extent defined for us by other persons. But there is still a core of personal being in 'self', (even if what is regarded as this core varies from social group to social group) which arises precisely because human agents <u>are</u> social beings, forged in action. For social beings must continually attempt to grasp what other persons mean (what they have 'in mind', so to speak). And to do this, they must also develop a facility for <u>self-monitoring</u> - one needs to recognise what one is doing oneself in order to make sense of and to others. But this is not action oriented towards explaining its own causal antecedents via theories (as in a 'science'), nor is it action motivated by some grand internal plan (as in cognitive theories). Rather persons are concerned to give narrations or <u>accounts</u> of what they are doing to themselves/to others (the two cannot be separated)(11) motivated by <u>intentions</u>.

There is a crucial difference between a theory and an account (Table 6).

Table 6: <u>Some Characteristics of Theories and</u>
<u>Accounts</u>

Theories	Accounts
retrospective	prospective
cognitive	perceptual
universal	Contextual
logical	rhetorical
abstract	metaphorical/examples
reasons	causes
reporting	telling
believing	knowing

A theory is concerned with taking a set of events that already existed prior to it and made one kind of sense and reshaping them to make quite another kind of sense. It is a cognitive operation. In contrast, an account is a perceptual operation, a more explicit description of what an action taking place within a particular everyday context actually <u>is</u>:

> The rhetorical essence of an account is that in its telling it serves a formative function, it works practically, to instruct or inform others as to how, in a sense, to be; it works to influence how they understand and experience things practically, i.e., in a direct and unmediated way; in other words, immediately and routinely. In its telling it is self-specifying in the sense of constructing or organising a setting or context within which its telling makes sense (Shotter 1985, p.40).

Of course, persons cannot account for all

their conduct, and nor do they need to. For it is often quite obvious to others what they are doing. It is only normal to request an account when it is not immediately obvious what a person is up to (for example, by reference to future aims, or by outlining what criteria they were trying to apply). But crucially, in making these accounts persons do not only explain themselves to other persons, they also explain themselves to themselves. Thus a person's account is also a process of self-monitoring through the medium of these accounts, a kind of internalised accounting, a working out of what one is doing. And as the ways of accounting for oneself change, so the person and the self can change.

Through their accounts, fashioned out of joint action, persons also construct the world beyond their immediate context. They learn progressively to specify regions of the world beyond themselves - school, the state, work, the economy(12). Often these accounts are loosely interlinked. When such linkages take place the resulting assemblage forms a 'folk model' (Holy and Stuchlik 1981, 1983), an account-based notion of what some piece of the world is like. Such folk models can be conjured up to justify particular actions by the person concerned or to explain the actions of other persons and social institutions. Thus, "I always vote Labour. But when they get into power they always let us down', or 'I shall vote Conservative again. With the Tories the small man has a chance to improve' or 'I'm not sure there is a God. If there is one, why should He allow all the terrible things that are happening', or 'Nothing much ordinary people can do' (Mass-Observation 1947). According to how well these folk models allow people to account for the world and for themselves, they invest people with a capacity for action. They are, therefore, the chief source of agency, the explicit formulation by a person of what she is capable of doing and of what powers that person has.

Finally, intentions (see Gauld and Shotter, 1977; Harré 1979, 1983; Shotter 1984). Just as accounts are not the same as theories, so intentions are not the same as plans. Rather, they are more like signposts. Intentions are usually vague and indeterminate, and by no means always directed to well-specified ends. Rather intentions consist of the progressive self-specification of projects already vaguely formulated, projects which change as they are more completely specified.

These, then, are the bare bones of the constructivist model of the human agent at the core of which, it should now have become clear, is the notion that man is the story-telling animal. It should also have become clear that the content of persons, selves, accounts and folk models must vary according to the contexts which have to be negotiated, contexts whose form will in turn depend upon the presence or absence of particular social institutions and the stocks of knowledge these institutions hold. Clearly, the exact extent of such variation is contingent on the precise nature of these institutions in particular places and times and must be empirically specified.

This is all well and good but how is it possible to get at persons' accounts and folk models, at how persons tell the world to themselves and themselves to the world? In the next section, I want to outline, in a preliminary way, how I am attempting to set about this task through a study I am carrying out of the well-documented 'leftwards' shift in political 'attitudes' or 'consciousness', that is in persons' accounts and folk models of politics, in England during the Second World War. Clearly, such an account must also involve consideration of other elements of a theory of social action, most especially the stocks of knowledge and 'the social' institutions from which the construction of persons, selves, accounts and folk models are inseparable (Thrift 1983b, 1985a).

A Better World. The Shift to the Left in English Political Accounts During the Second World War

> I feel quite exhausted after seeing and hearing so much sadness, sorrow, heroism and magnificent spirit. The destruction is so awful and the people so wonderful - they deserve a better world (Queen Elizabeth to Queen Mary in a letter, 19.10.1940, cited in Calder 1969, p.605).

> Promises of a New World won't help us. (Mass-Observation 1947, p.322).

It is important to avoid exaggerating the extent of the shift to the left(14) in England during the Second World War leading to the Labour victory in the 1945 General Election, especially since the whole episode has taken on something of the status of a latter-day myth amongst the left - the build-up of working class aspirations, the great victory and then the betrayal (see Stedman Jones 1983; Pimlott 1985). Politics occupies very little of the accounting of most persons' lives and even the extreme circumstances of the Second World War did not lead everyone to reconsider their accounts and revise them (indeed, for many it simply confirmed what they already 'knew'). Thus the emphasis in this section is on the fairly large minority whose accounts did change in some way during the Second World War(15). Even this minority constituted a quite remarkable shift - at least in the context of British politics - more particularly because it came about in the absence of much in the way of party organisation (on left or right) and was therefore relatively spontaneous. In what follows, in order to restrict the example to a manageable length, I have concentrated on changes on the Home Front. Although it has become something of a truism that it was the Forces vote that accounted for the 1945 Labour victory, in point of fact although in June 1945 there were 4,531,000 men and women in the Forces over 21 years of age, only 1, 701,000 were able to cast votes' (Calder, 1969, p.671). 'In total it did not amount to the number of votes separating the two major parties, let alone account for the strong swing of the pendulum' (Harrington and Young, 1978, p. 202). Rather a considerable number on the Home Front had also changed their accounts so that the Home Front is a valid focus of interest.

The 1945 Election provides a very rough guide to who these people were and where they lived. Table 7 outlines the 1945 General Election results (see the analyses in McCallum and Readman, 1947; Butler and Stokes, 1969 and Eatwell, 1979). Between the 1935 and 1945 elections there was a twelve per cent swing (in both England and the UK as a whole) to Labour and the associated parties on the left (Common Wealth, I.L.P., Communists) from the Conservatives and their associates on the right (the National Liberals, the National Party and various Independent Conservatives) (McCallum and Readman 1947). Labour won 393 seats, up from 154 in 1935(16). The Labour vote increased from 8,325,491 in 1935 to 11,992,292 in 1945; so more than three and a half million more people voted Labour.

Studies of the social composition of the Labour

Table 7:

The General Elections of 1935 and 1945

	per cent electorate voting	per cent Conservative, Liberal, Independent Conservative, National	per cent Liberal	per cent Labour, ILP, Commonwealth, Communist	others
England					
1935	71	55	21	41	14
1945	73	40	17	50	9
Great Britain and Northern Ireland					
1935	71	55	24	41	20
1945	73	40	18	50	11

Source: MacCallum and Readman (1947)

vote showed that this increase relied upon three
inter-related groups. First, there was considerable
support from the young; Labour won 61 per cent
of the new votes in 1945 (Eatwell 1979).   (This
support continued on; according to Butler and Stokes
(1969) people who were young in the 1940s were
of all age groups most prone to vote Labour in
the 1960s.) Second, there was evidence of a limited
swing to Labour amongst the middle classes. 'Polls
showed that Labour received 21 per cent of the
middle class vote, almost certainly more than ever
before, though well below the Conservatives 54
per cent (Eatwell 1979, p. 42).   Some of this vote
came from reforming idealists but much of it could
be accounted for by the growth of 'white collar'
office workers (Bonham, 1954).   Butler and Stokes
(1969), for example, found that the 1935-1950 cohort
of electors contained seven per cent more middle-
class voters, most of whom were white-collar office
workers.   Third, and most important of all, Labour
gained the working-class vote, probably as much
as 70 per cent of it (Calder 1969).   Indeed, more
than ever before, the Labour vote was a class vote.
Mass-Observation found that by far the main reason
given for voting Labour was 'class identity'; 43
per cent gave this as their reason, compared with
six per cent for nationalisation, the next specific
reason.
    The electoral geography of Britain changed
in line with the social composition of the vote
(Figure 3 and 4)(17). The most strong Labour gains
were in three places. First, there were the London
suburbs where Labour won many 'white collar' constituencies. Second, there was the West Midlands;
before 1945 the traditional home of the working-
class Conservative.   Finally, there were the
agricultural counties of East Anglia, where many
agricultural labourers voted Labour. To summarise,
in the 1945 election, 'the working class of the
declining areas of heavy industry joined forces
through the ballot box with much of the more prosperous working class in the Midlands and South
East, and a substantial section of the urban middle
classes' (Addison, 1975; p.268).
    But how did changes in political accounts
and folk models come about in the Second World
War that laid the foundations for the Labour victory?   Conventionally, a number of reasons are
put forward.   The first, and the most often cited,
is the amount of left-wing 'propaganda' that was
being disseminated, compared with right-wing

Figure 3: The Geography of the 1935 General Election

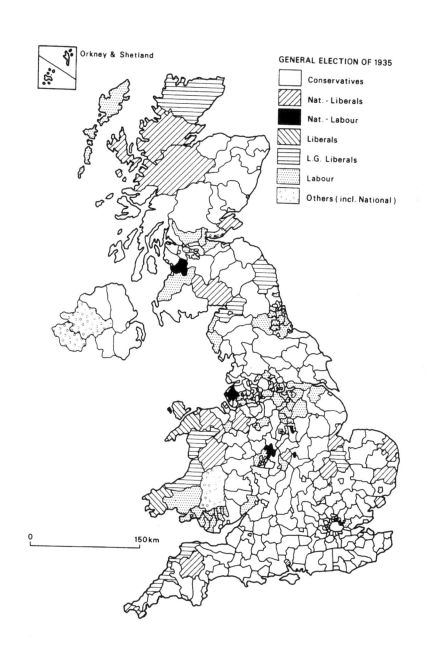

GENERAL ELECTION OF 1935

Conservatives

Nat. - Liberals

Nat. - Labour

Liberals

L.G. Liberals

Labour

Others (incl. National)

Orkney & Shetland

0   150 km

Figure 4: The Geography of the 1945 General Election

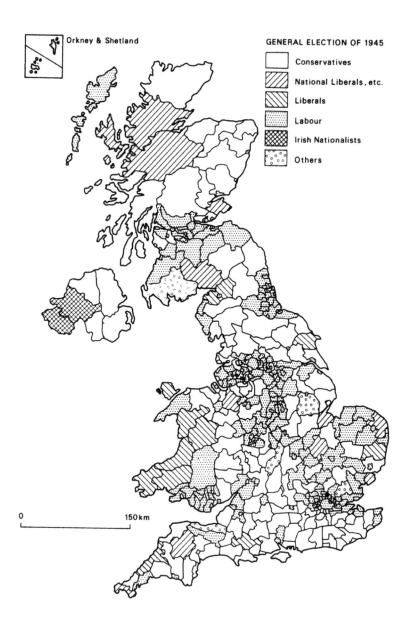

material. Certainly, the need for the creation
of a new Britain was strongly pressed in newspapers
like the Daily Herald, the Daily Mirror, even The
Times and in magazines like Picture Post. Pam-
phlets, periodicals and most particularly the
Penguin Specials helped the process along. Some
programmes on the radio had a pronounced left-
wing bias or, as in the case of J.B. Priestley's
'Postscripts', gave voice to a general feeling
of discontent. Films were made that scored left-
wing political points. The Church of England,
in contrast to its loyal jingoism in the First
World War, showed distinct signs of independence
embodied in the person of Archbishop Temple of
Canterbury. Even the Government, to some extent,
pushed the process along through praise for the
Soviet Union at particular junctures in the
War(18). But, as has been pointed out by a number
of writers (e.g. Calder, 1969; Harrington and Young,
1975; Pelling, 1984), just because such 'propaganda'
was about does not mean that people were automatically receptive to it; indeed, the evidence
is that outright propaganda aroused antagonism.
More likely is it that the availability of left-
wing material tended to reinforce existing views
and tendencies so that:

> it is reasonable to doubt the extent to which
> it produced socialists in any serious sense
> of the word, especially as much of the
> propaganda was compatible with moderate
> reformism. If anything, the propaganda served
> more to reinforce Labour's attack on the record
> of the inter-war Conservatives. Reformist
> and socialist propaganda also had to be seen
> against a counter-cultural background, and
> one in this case which stressed strong tra-
> ditional values. For example, the majority
> of literature produced during the war, let
> alone read, had far from socialist implications
> or values (Eatwell, 1979, p.39).

A second and probably more salient reason
for the shift to the left in accounts and folk
models - because it was firmly grounded in people's
lives - was that genuine economic changes came
about during World War II and, most especially
the creation of full employment(19) and significant
wage rises amongst the working class so that, even
given considerable increases in the cost of living,
earnings were substantial (although much of the

increase in working class prosperity could be accounted for by the massive amount of overtime that was worked rather than dramatic rises in the hourly earnings paid). 'There is no doubt that large sections of the working class were better off as a result of the war, and that there was levelling up, as well as levelling down, towards the skilled artisan's standards' (Calder 1969, p.63). Clearly, there was concern amongst the working class that full employment and economic gains should not be eroded after the war (when so many in the Armed Forces would be demobilised) and the Party which had secured the credit for making a stand in favour of the Beveridge Report was clearly the one most likely to protect them.

A third reason, one that clearly had influence on the formulation of some people's accounts and folk models, was the practical demonstration brought about by the massive State intervention in the direction of industry during the War (see Ministry of Labour and National Service 1947) that nationalisation was a feasible course to take. This reason combined, to some extent, with the general popularity of Russia.

A fourth reason for left wing changes in accounts and folk models was the general desire for better housing and welfare. The Gallup Polls show housing, not surprisingly given bomb damage, relocation and cessation of building, to be a continual preoccupation through the course of the War. 'Housing was for most electors the most important social issue...Labour candidates stressed the importance of this issue and many of them had served on the housing committees of their local Councils which dealt directly with the problem' (Pelling 1984, p.32).

Finally, there was the 'swing of the pendulum'. There had been a Parliament with a Conservative majority since November 1931 and discontent had accumulated against it.

But the problem is that people's accounts and folk models are not constructed as a disembodied set of 'reasons', all neatly ordered. They are situated in, and adjusted to, particular contexts. They are <u>active</u> talk, not passive contemplations. They are set 'within the socially organised worlds in which they participate as constituting and constitutive elements' (Heritage 1984, p.178). People's accounts and folk models are contextual and what is certain about the Second World War is that a number of these contexts changed very dramatically indeed.

The Contexts of Talk

Perhaps the main reason why the shift to the left was so marked was the substantial number of changes that took place in the intricate mosaic of contexts that went to make up English society. Those changes were sufficient to force some people to re-account for their world, either by virtue of changes to the contexts that they knew or by catapulting them into contexts to which they had never previously been exposed (and to which they had to try to adjust). In this joint talk, people were therefore able to reach different conclusions from those that they had come to before the War. There were many changes to the mosaic of contexts; here I will concentrate on just two. The first was the enormous amount of movement that took place in World War II. The second was the growth of new institutions of sociability and the transformation of a number of the old institutions.

The amount of movement that took place in World War II, and with it new patterns of social interaction, was considerable. There were three main sources of movement. The first of these was evacuation from the urban areas most vulnerable to bombing to rural 'reception' areas. By 1939, some two million people (mostly those with some means) had evacuated independently and another million-and-a-half (mainly poorer) people had been evacuated in the Government scheme (see Mass-Observation 1940; Padley and Cole 1940; Titmuss 1950) (Figure 5). There was a great difference in response (Table 8) - under half of London's schoolchildren went but over 70 per cent of Newcastle and Gateshead's. The 1939 evacuation was often chaotic and suffered from a strong drift back to the cities when the expected bombing did not immediately materialise but this and the subsequent evacuations (particularly at the height of the bombing in 1941 and during the V1 attacks in 1944) had important social effects, intimately exposing people from quite different classes to one another for the first time.

The second source of movement, and probably the one with the greatest effects, stemmed from employment policy. The composition of the labour force changed dramatically during the Second World War (Table 9) (See Fogarty 1945; Ministry of Labour and National Service 1947; Planning 1947). From

Figure 5: Total Numbers Billetted in Reception
Areas Under the Official Scheme

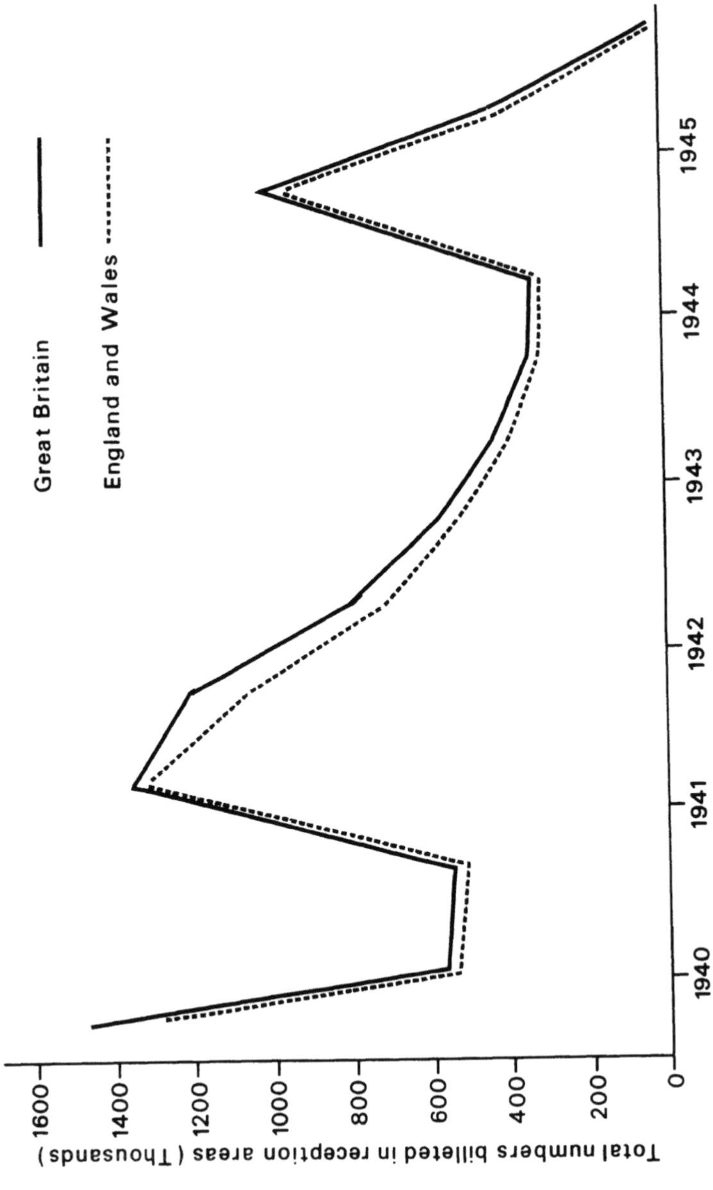

Great Britain

England and Wales ....

1600   1400   1200   1000   800   600   400   200   0

Total numbers billeted in reception areas (Thousands)

1940   1941   1942   1943   1944   1945

Table 8: **Proportion of unaccompanied schoolchildren and mothers and children evacuated from some of the major urban areas of England at the outbreak of war**

County borough	Per cent of children evacuated	Per cent of mothers and children evacuated
Newcastle	71	57
Gateshead	71	49
South Shields	31	26
Tynemouth	32	29
Sunderland	33	20
West Hartlepool	36	24
Middlesbrough	31	18
Leeds	33	26
Bradford	25	17
Bootle	68	66
Liverpool	61	44
Wallasey	76	–
Birkenhead	62	44
Manchester	69	44
Salford	76	56
Rotherham	8	6
Sheffield	15	13
Derby	27	14
Nottingham	22	12
Walsall	18	)
West Bromwich	26	) 11
Smethwick	24	15
Birmingham	25	21
Coventry	20	14
Portsmouth	30	20
Southampton	37	28
London Administrative County	49	–
County Boroughs and London Administrative County	46	37

Source: Titmuss (1950), pp. 550-552.

Table 9:   Employment in Great Britain, 1938 and 1944 (thousands)

Category of employment	Males aged 14-64				Females aged 14-59(1)				Total			
	Number		Per cent		Number		Per cent		Number		Per cent	
	June 1939	June 1944	June 1939	June 1944	June 1939	June 1944	June 1939	June 1944	June 1939	June 1944	June 1939	June 1944
Armed Forces and Women's Auxiliary Services	477	4,502	3.0	28.3	-	467	-	2.9	477	4,968	1.5	15.6
Civil Defence	80	225	0.5	1.4	-	56	-	0.3	80	281	0.2	0.9
Group I (munitions)												
Metal and Chemical Industries(2)	2,600	3,210	16.2	20.2	506	1,851	3.1	11.6	3,106	5,061	9.7	15.9
Group II (essential services)												
Agriculture, horti-culture	1,046	948	6.5	6.0	67	184	0.4	1.1	1,113	1,132	3.5	3.5
Mining	868	802	5.4	5.0	5	13	-	0.1	873	815	2.7	2.5
National Government Service	416	520	2.6	3.3	123	495	0.8	3.1	539	1,015	1.7	3.2
Local Government Service	520	322	3.2	2.0	326	468	2.0	2.9	846	790	2.6	2.5
Gas, Water and Electricity Supply	225	160	1.4	1.0	17	32	0.1	0.2	242	192	0.8	0.6
Transport, Shipping and Fishing	1,222	1,038	7.6	6.5	51	212	0.3	1.3	1,273	1,250	4.0	3.9
Food, Drink and Tobacco	391	269	2.5	1.7	263	240	1.7	1.5	654	509	2.0	1.6
Total, Group II	4,688	4,059	29.2	25.5	852	1,644	5.3	10.2	5,540	5,703	17.3	17.8
Group III (less essential services)												
Building and Civil Engineering	1,294	600	8.1	3.8	16	23	0.1	0.1	1,310	623	4.1	2.0
Textiles	401	221	2.5	1.4	601	405	3.8	2.5	1,002	626	3.1	2.0
Clothing/Boots	138	65	0.9	0.4	449	284	2.8	1.8	587	349	1.8	1.1

Table 9 (Continued)

Category of employment	Males aged 14-64				Females aged 14-59 (1)				Total			
	Number		Per cent		Number		Per cent		Number		Per cent	
	June 1939	June 1944	June 1939	June 1944	June 1939	June 1944	June 1939	June 1944	June 1939	June 1944	June 1939	June 1944
Group III continued												
Boots and shoes	108	64	0.7	0.4	57	43	0.4	0.3	165	107	0.5	0.3
Other Manufactures (3)	1,004	542	6.2	3.4	440	414	2.7	2.6	1,444	956	4.5	3.0
Distributive Trades	1,888	972	11.8	6.1	999	956	6.2	6.0	2,887	1,928	9.0	9.0
Other Services (4)	965	436	6.0	2.7	917	977	5.7	6.1	1,882	1,413	5.9	4.4
Total, Group III	5,798	2,900	32.2	18.2	3,479	3,102	21.7	19.4	9,277	6,002	29.0	18.8
Total of Armed Forces, Auxiliary Services, Civil Defence and Industry	13,643	14,896	85.1	93.6	4,837	7,120	30.1	44.4	18,480	22,016	57.7	69.0
Unemployed	1,043	71	6.5	0.4	302	31	1.9	0.2	1,345	102	4.2	0.3
Rest of population	1,324	943	8.4	6.0	10,901*	8,869*	68.0	55.4	12,225	9,812	38.1	30.7
TOTAL	16,010	15,910	100.0	100.0	16,040	16,020	100.0	100.0	32,050	31,930	100.0	100.0

Notes (1) Women working part-time are included throughout, two being counted as one unit. At the middle
of 1944 about 900,000 women were doing part-time work.

(2) Metal manufacture, engineering, motors, aircraft and other vehicles, shipbuilding and ship-
repairing, metal goods manufacture, chemicals, explosives, oil, etc., industries.

(3) Leather, wood, paper, bricks, tiles, pottery, glass and miscellaneous manufactures.

(4) Commerce, banking, insurance, finance; professional services; entertainment; hotels,
restaurants, etc..; laundries and cleaning.

(*) Mainly housewives. Domestic servants are also included.

Source: Fogarty (1945), pp. 46-47.

LITTLE GAMES AND BIG STORIES:

June 1939 to June 1944 some four-and-a-half million people were added to the Armed Forces, two million to the munitions industries and between a quarter and a half-a-million to Civil Defence and the essential industries.

> Of this total, 3½ million were found by transference from less essential industries and services, the largest single contribution coming from the distributive trades (almost a million) and building. A million-and-a-quarter unemployed were brought back to work and the balance was made up with nearly 2½ million people not normally engaged in paid work other than domestic service, including two million women. In all, the number of workers in the less essential (Group III) industries was reduced by 35 per cent between 1939 and 1944, and 20 per cent of the 'rest of the population' were transferred into war work of one kind or another. If each part-time worker were counted separately the latter proportion would be around 23 per cent (Fogarty 1945, p.48).

The result of this reorganisation was a much longer journey to work for many workers and, more particularly, numbers of workers were directed to move to new areas where they were then billetted in hostels, lodging houses and private homes. Some smaller towns received large influxes of population (Table 10) and suffered from severe overcrowding as a result(20).

A special effect of the wartime employment policy was the enormous number of women who went to work. Many went voluntarily but from 1942 'conscription' operated (Mass-Observation 1942; Minns 1980). Some 80,000 women joined the Land Army over the course of the War; 100,000 women worked on the railways, 300,000 women joined the chemical industries; and another one-and-a-half million women went to the engineering and metal industries (Table 9). Most women worked relatively near their homes but those unmarried and over 19 years of age and less than about 50 years were declared 'mobile' and likely to be moved considerable distances (Ministry of Labour and National Service 1947).

The final source of movement stemmed from the direct effects of the War itself, especially the bombing. In paricular, many people were made

Table 10:

Population increases in selected towns whose population increased by four per cent or more, 1938-1942

Town	Population in October 1942	Percentage increase in population:	
		mid-1938 to October 1942	mid-1938 to June 1942
Reading	115,193	19.7	28.6
Slough	63,712	25.9	21.1
Cheltenham	60,488	16.5	15.4
Swindon	67,862	12.0	9.6
Blackpool	152,128	20.9	15.3
Southport	88,708	12.9	4.9
Luton	100,480	10.6	9.6
Romford	60,487	10.8	9.8
Watford	68,789	4.7	1.9
Northampton	100,502	4.1	1.9
Enfield	95,579	4.0	2.9
Harrow	193,745	5.6	3.4

Source: Combined Production and Resources Board (1945), p.103.

homeless (Table 11). At least one in six London households (or 1,400,000 people) faced homelessness at some time during the war. In Plymouth the figure was one in four (Calder 1969).

Table 11: Housing out of civilian use in the UK, 1944

	mid – 1944	end – 1944
Houses destroyed or damaged beyond repair by enemy action	175,000	205,000
Houses damaged, uninhabitable and awaiting repair	80,000	100,000
Houses requisitioned for military use or otherwise withdrawn	70,000	–
Houses evacuated in south-eastern coastal areas	125,000	–

Source: Combined Production and Resources Board (1945), p.42.

The magnitude of these three movements can be seen, in net terms, from the redistribution of population during the war. In particular, there was a net shift of between one-and-a-half and one-and-three-quarter million people from London, the Southeast and the East Coast regions towards the North and West between 1938 and 1942. The full pattern of inter-regional shifts is shown in Table 12. This Table does not, of course, take into account intra-regional population movements, but:

the most striking changes took place for the most part within regions, without affecting the inter-regional balance. In the towns and districts which have been most heavily hit by evacuation, or in areas such as the belt of counties to the North and West of London, or the counties of North Wales, or the parts of Western Scotland which are not yet industrialised, there have been changes very much greater than any recorded (inter-regional changes) (Fogarty 1945)(21).

The second change that made some people re-assess their world and themselves was the enormous growth of institutions of sociability. These institutions can be divided into two types, those promoted by the State (and other compulsory) and voluntary organisations.

Amongst the new State institutions of sociability the most notable were the Home Guard, the Royal Observer Corps and Civil Defence (that is, air-raid wardens, rescue and first-aid parties, report and control centre staff and messengers). The Home Guard, originally formed as the Local Defence Volunteers in 1939, employed the talents of one-and-a-half million people during most of the War (see Figure 6) mainly part-time. The Royal Observer Corps, founded in the 1920s, also had a large complement of part-timers and set to work over 30,000 in its ranks during the rest of the War (Figure 7). Finally, there was the whole group of Civil Defence occupations (Figure 8). Prominent amongst these were the air-raid wardens. There were some 200 to 250,000 of these in London alone (Calder 1969).

There is little systematic information about the social composition of the Home Guard, the Royal Observer Corps and Civil Defence. In the Home Guard there were few women, in the Royal Observer Corps the proportion was one in eight, in Civil Defence the proportion was one in six. Social composition varied strongly with the social composition of the neighbourhood. For example, in many boroughs of London, as might be expected, two thirds of air-raid wardens were working class, but in areas outside London the proportion was usually less (Calder 1969). Certainly, however, the three new institutions were not, given the pronounced class nature of British society at the time, as strongly class-divided as might have been thought. For example, the Home Guard

was not just a regrouping of the British Legion. In the villages, poachers and game-keepers, farmers and farm hands, marched to-gether. In the mining districts, the colliery

Table 12:   The distribution of population of Great Britain by region, 1938 and 1942(1)

Region	Percentage of population living in each region		Actual estimated population	Estimated increase or decrease in the civilian population, 1938-1942, allowing for the change in the civil population in the country as whole(2)	
	June 1938 total population	April 1942 civilian population	June 1938 000's	Number 000's	Per cent of 1938
London (LCC and Middlesex)	13.3	10.3	6,121	-1,308	-21.3
Eastern (Hertford, Bedford, Essex, Cambridge, Ely, Huntingdon, Norfolk, Suffolk)	8.4	8.1	3,858	- 95	- 2.4
South-eastern (Kent, Surrey, Sussex)	7.9	7.6	3,670	- 153	- 4.1
Southern (Oxford, Berkshire, Buckingham, Hampshire, Isle of Wight, Dorset)	4.9	5.3	2,286	+ 170	+ 7.5
South-western (Wiltshire, Gloucester, Somerset, Devon, Cornwall)	5.7	6.7	2,646	+ 406	+15.4

Table 12 (cont.) <u>The distribution of population of Great Britain by region, 1938 and 1942(1)</u>

Region	Percentage of population living in each region		Actual estimated population	Estimated increase or decrease in the civilian population, 1938-1942, allowing for the change in the civil population in the country as whole(2)	
	June 1938 total population	April 1942 civilian population	June 1938 000's	Number 000's	Per cent of 1938
West Midlands (Warwick, Stafford, Shropshire, Worcester, Hereford)	8.5	9.1	3,935	+ 269	+ 6.9
North Midlands (North-ampton, Peterborough, Rutland, Leicester, Lincoln, Nottingham, Derby)	6.6	7.2	3,050	+ 280	+ 9.2
North-western, (Cheshire, Lancashire, Cumberland, Westmorland)	14.2	14.4	6,541	+ 91	+ 1.2
North-eastern (East and West Ridings: York)	8.6	8.4	3,964	- 77	- 1.9
Northern (Durham, North-umberland, North Riding)	5.8	5.9	2,677	+ 63	+ 2.4
Wales	5.3	5.9	2,466	+ 226	+ 9.2
Scotland	10.8	11.1	4,993	+ 128	+ 2.6
TOTAL	100.0	100.0	46,208	(+1,633) (-1,633)	-

Notes: (1) The population figures are estimates only.
(2) Calculated given that the total civilian population was 5.7 per cent smaller in 1947 than in 1938.

Source: Fogarty (1945), p.42.

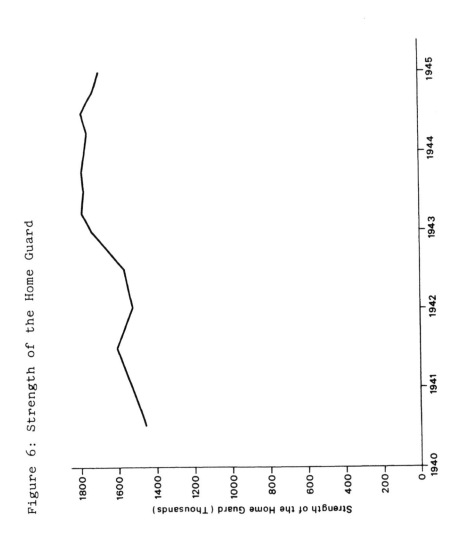

Figure 6: Strength of the Home Guard

Figure 7: Stength of the Royal Observer Corps

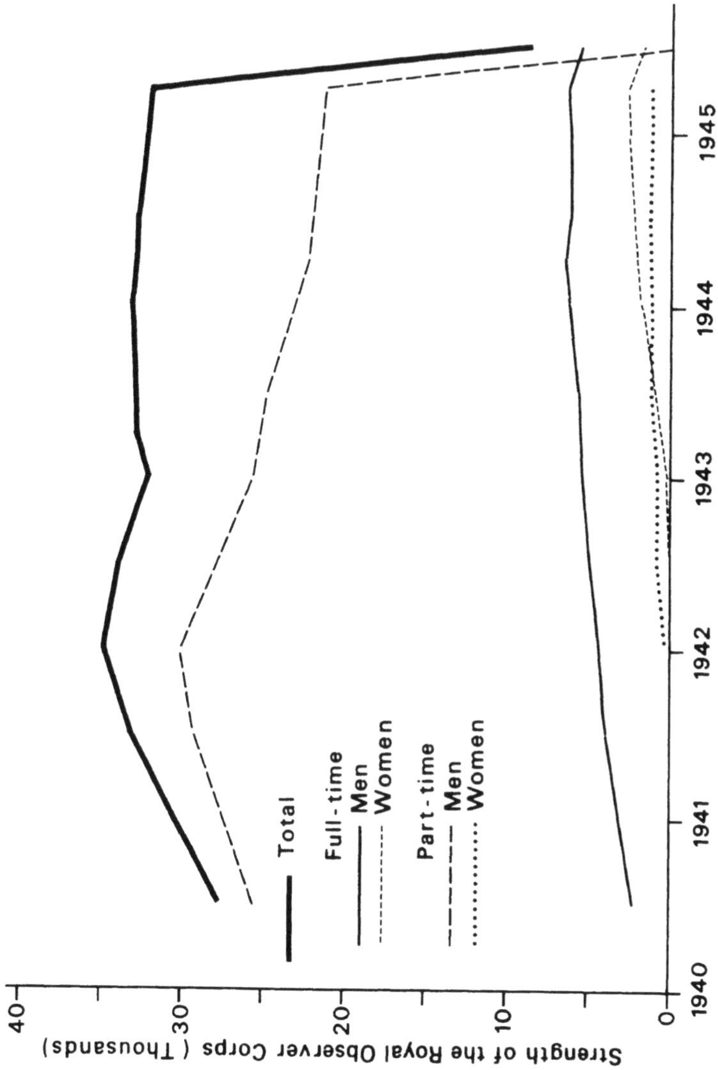

Figure 8: Numbers in the Civil Defence Services
of Great Britain 1940-1945

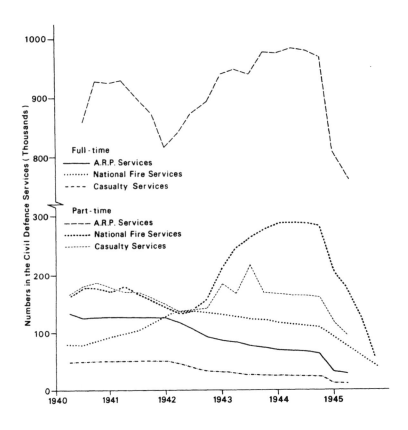

> L.D.V.'s would parade "in their dirt" soon after coming off a shift. Southern Railways set a lead for public utility companies by organising its own L.D.V. (Calder, 1969, p.193).

Further, the three institutions were not just semi-military organisations. They performed numbers of social functions (which further promoted the mixing of their members), from dances to darts matches, from football leagues to whist drives.

Many new voluntary organisations also grew up. Amongst the most important was the Women's Voluntary Service, founded in 1938. The exact numbers in the Service are unknown but certainly numbered in the 'hundreds of thousands' (Calder 1969; Minns 1980). Most of the women involved were old - over the 'mobile' age limit - and middle class (most working-class women did not have the time to belong), yet

> such was the onslaught of the...War that the W.V.S., <u>in certain times and places</u>, transcended its apparently basic middle classness. A post-war observer has revealed that before 1939 it "would have been inconceivable that in a philanthropic organisation the relatives of e.g. committee members and clients could be interchangeable or that members of particular trades or industries or even residents in a "poor" locality could take part in voluntary services..." But war disrupted the lives of middle-class and working-class women indiscriminately (Calder 1969, p.224).

Apart from new organisations, many existing institutions also grew during the War. In particular, trade unionism became a much more important force. Overall union membership in Britain increased from 6,053,000 at the end of 1938 to 7,803,000 by the end of 1945 (Ministry of Labour and National Service 1947). The unions involved in the expanding war effort obviously registered the largest increases. The Amalgamated Engineering Union, for example, had 413,000 members in 1939. But by 1943 membership had more than doubled to 909,000 (Calder, 1969). The boost in union membership was, of course, helped along by full employment, and, in many cases, severe labour shortages. Government intervention also contributed. By 1943, 4,500 joint production committees had been set

up which all required the presence of unions in
factories(22). However, the Unions were not the
only existing institutions to receive a boost from
the War. For example, existing voluntary organisations like the Women's Institutes and the Red
Cross also grew(23).

These changes in spatial and social organisation meant that the contexts within which new
information, economic prosperity and the other
likely determinants of change in people's accounts
had an impact were radically different from before
the War. Thus, England in the Second World War
had become a society over which a new 'developmental
matrix' (Shotter 1984) that included elements of
many new contexts, had been laid, a matrix in which
spaces were opened up in which some persons could
live and develop in ways that previously would
have been unavailable to them. These persons could
account for the world in different ways and create
new folk models of politics. They might even become
different persons.

The New Accounts

In this final section of the paper I want to consider the changed accounts and folk models of the
Second World War which lead to the Labour victory
in the 1945 election. It is natural in these circumstances to turn to the work of Mass-Observation,
that unique organisation founded in 1937 which
- amongst other things - collected accounts throughout the course of the War and billed itself as
creating 'the science of ourselves'. I am currently
involved in a study aimed at reconstructing the
multiple contexts of two of Mass-Observation's
key locations - 'Worktown' (Bolton) and 'Metrop'
(Fulham) with a view to eliciting how working class
accounts and folk models could and did change during
the Second World War. In what follows I provide,
in lieu of this specific study, a very general
illustrative survey stretching across different
contexts drawn from all classes, with a short election postscript from Fulham. I will illustrate
the survey with snippets of accountings but it
is vitally important to stress that these snippets
are but fragments of an ongoing process of talk
developing out of particular contexts. These
snippets are not (although out of context, they
may appear to be) 'reasons', finished contemplative
conclusions. They might well have changed in an
other context. The ebb and flow and the kind of

talk from which these accounts are drawn and from which they are separated at such peril, is well illustrated by the following report on the aftermath of the 1945 Election.

> We got a real shock when we heard our Conservative member had been beaten by 12,000 - we simply could not believe it. Then, as more women came in to sew, it was 'like a jam session', everyone talked at once, and I gave up trying to sew or give out work, and sat round talking - or, more correctly, listening. The general opinion was that the 'soldiers' vote' had swung the Election. Some thought that England would go down in the world's opinion, that no one would believe in her stability, and that Russia would rejoice since it would <u>suit</u> Stalin. Someone wondered if we could get the road across Morecambe Bay now, as Labour 'doesn't care about spending money or consider whether it is practical': 'Now the coal mines <u>will</u> be nationalised - see what they make of <u>that</u>'. I heard remarks about soldiers 'going solidly for Labour' - of 'agents' in the Army who had plugged Labour views. The Tories, the big businessmen, had always made the war in the first place - it was the big men who put Hitler into power, who sold armaments to <u>any</u> nation whether they were going to use them against our friendly nations or not, that it was the Tory plan to send girls away from home to work, send boys into the mines, dole out rations barely enough to keep people alive. Any stick seemed good or bad enough to beat the 'Tory dog'. Some soldiers seemed to think the Government had fallen down on the housing plan - and would do so on the demob plan. One woman told of a lad who voted Labour because he <u>knew</u> he would get out quicker to his father's butcher business (Broad and Fleming 1981, p. 296).

Out of the very many ways in which the 'developmental matrix' of contexts which people lived shifted during the Second World War, three ways are of particular importance. The first of these was the generally increased level of sociability, promoted by both movement and the new institutions of sociability. A much greater degree of interaction was forced upon people, often for the first time in their lives, and this interaction was not just

within classes but between classes. Two examples
of the impacts of this interaction on accounts
will suffice. One example, the first wave of
Government sponsored evacuations, springs to mind
as a particularly graphic example of this phenomenon. 'Social mismatching was inherent in the
scheme' (Calder 1969, p.46). For the first time
many working-class parents and children who were
billetted with middle-class families got to see
how the other half lived while the middle class
got to see how the working-class parents and
children lived(24). The shock was great on both
sides (although it has been exaggerated since).
Here were children who never wore pyjamas(25),
who had few clothes, who suffered from lice, scabies
and impetigo and who, worst of all, in five to
ten per cent of cases lacked proper toilet training.
On both sides prejudices were confirmed. Indeed,
rather interestingly, on both sides the accounts
are nearly all made in class terms. For example,
from the working class:

> 'Think it was bloody Buckingham Palace the
> way she goes on;' (Mass-Observation 1940,
> p.319).
> 'It's too posh for me' (Ibid., p.314).
> 'We'd rather have our meals in the Kitchen'
> (Ibid., p.330).
> 'Two children from a very poor house were
> taken care of by very nice people who were
> teaching them to use table napkins and such
> like things that the children had never seen
> before. The mother came to see them and com-
> plained that they were being brought up 'narky'
> and that she wasn't going to have it' (Ibid.,
> p.314).

or most to the point:

> 'You can't be yourself here' (Ibid., p.314).

And from the middle class, the reaction was not
much different:

> 'The butler has been disgruntled ever since
> I told hom we were having a number of children,
> and consequently would start living more simply
> ourselves - stopping entertaining and so forth'
> (Ibid., p. 329).

Mass-Observation reported of the middle class,
with some degree of hyperbole:

> A minority, mainly represented in the women's
> organisations have turned horror into pity,

and has determined that the appalling conditions which the evacuee children reflect shall
be swept away, and very soon; the majority
have turned their horror into fear and even
hatred, seeing in this level of humanity an
animal threat, that vague and horrid revolution
which lurks in the dreams of so many supertax
payers (Ibid., pp. 336-337).

Another example of the generally heightened
level of interaction was, of course, the number
of people - particularly women - who went to work
for the first time or went to work in quite different locations. In the factories and workshops
there was considerable forced mixing of people
from quite different backgrounds which showed up
in accounts of the time. Take the case of women
in production (Mass-Observation, 1942). Not only
working-class women were mobilised. One young
teacher who helped make anti-tank mines in her
summer holiday noted in her diary.

Chief topic of conversation the Dizzy Blonde.
She arrived last Monday and has caused quite
a sensation, partly by her appearance - wonder-
ful peroxide curls, exquisite make-up, etc.
- but much more by her conversation. Has told
everyone that she has never worked before
as she has an allowance and is going to frame
the first pound she gets. Another tale runs
that she lost a wallet with £100 in but "just
didn't bother"; she's saving up for a fur
coat, a really good one; "I had 5 but I've
given 3 away". Mrs. Barratt's comment, "Now
we know what she did before the war" (Mass-
Observation Diary quoted in Calder 1969, p.
385).

Many women tasted independence for the first time:

When you get up in the morning you feel you
go out with something in your bag, and something coming in at the end of the week, and
it's nice. It's a taste of independence,
and you feel a lot happier for it. (File Report
No. 2059 quoted in Calder and Sheridan 1984,
p.179).

But only the young seemed willing to continue their
independent lifestyle after the war was over:

> I'd like a real change after the war, but
> I'll have to see what jobs are going. I think
> a lot will leave this country after the war
> - those that have married Americans and
> Canadians, and those that have heard what
> it's like over there - I know a lot would
> like to get right out of this country, they
> think there'll be more freedom over in America
> and Canada. I should think a lot more will
> want to go than can get. I'd like to myself.
> (File Report No.2059 quoted in Calder and
> Sheridan 1984, p.180).

Most worries for the future seemed to centre
on employment after the War.

> We shall be kicked out (Mass-Observation 1942,
> p.328).
> There won't be jobs for people like myself
> (Ibid., p.328).
> There won't be jobs for women (Ibid., p.328).

> I think there'll really be a lot of unemploy-
> ment - look at all the girls in the services
> coming back and wanting jobs, to say nothing
> of the men. They can stay in the army for
> a time, but the girls can't, I think we're
> in for a very difficult time (File Report
> No. 2059, cited in Calder and Sheridan 1984,
> p.182).

> It's not so much what's going to happen to
> us, as what's going to happen to the men who
> come home. Will there be jobs for them? (File
> Report No. 2059, quoted in Calder and Sheridan
> 1984, p.182).

> I think there ought to be plenty of work for
> everybody after the war, if they manage it
> properly (Ibid., p.182).

The second, closely inter-related, way in
which changes in contexts came about was through
an increase in the stocks of knowledge that were
available and through changes in what constituted
these stocks of knowledge. Stocks of knowledge
were greatly increased. Ways of doing things that
had been considered inadmissible before the war
became acceptable behaviour. Availability of printed knowledge was probably much greater than before
the war and was likely to be taken advantage of

(certain jobs like firewatching promoted a reading habit - there was little else to do). Book sales went up by 50 per cent. Newspaper sales increased. Only magazine sales decreased and this because numerous magazines were no longer printed (Table 13). Much of the increase in book sales can be accounted for by mundane facts - over a million volumes were lost from public libraries in bombing and had to be replaced (Calder, 1969) - but still even this evidence suggests that the reading habit had spread to new sections of the population. And, of course, such evidence says nothing of the greatly increased incidence of multiple readership of single volumes in the Second World War - books were passed around(26).

It seems that the effect of hearing more about the world that wartime experiences and increased reading brought about had some effect on some people's accounts, an effect that was cumulative. It was on these people that left-wing material had most effect. Here is a young Clive Jenkins:

> There were lots of private libraries in Wales and the manager of our local Co-op Insurance lent me books. He gave me an Introduction to Marx. The full-time secretary of the Band of Hope and Temperance Union was a social democrat and lent me (Left) Book Club choices like Attlee's The Labour Party in Perspective, which was really a pretty boring old book. And then I got introduced to things like Guilty Men and The Trial of Mussolini. I remember you had to put your name down on the waiting list for your copy of Guilty Men (quoted in Harrington and Young 1975, p.66).

Certainly, the demand for 'serious' books (and radio programmes like The Brains Trust) concerned with current affairs seems to have increased. For example, Penguin Specials had an average sale of 80,000 copies each and books like Guilty Men sold many more. When the Beveridge report (Cmd 6404) on social security was published in December 1942, it sold 635,000 copies in full or summary form(27). People queued outside the HMSO bookshops for it and, within two weeks of publication, a Gallup Poll found that 19 out of 20 people had heard of the report (Calder 1969).

By the time of the 1945 Election, Richard Crossman could write that there had been:

Table 13 Consumer purchases of newspapers, magazines and books in the U.K.

	1938	1939	1940	1941	1942	1943	1944
Total value at current prices (£ million)	62	61	62	70	73	77	78
Total value at 1938 prices (£ million)	62	61	57	58	60	63	64
Per capita value at 1938 prices (£)							
(a) Total	1.31	1.28	1.19	1.22	1.26	1.33	1.33
(b) Newspapers and magazines	1.10	–	–	–	–	1.00	1.01
(c) Books	0.21	–	–	–	–	0.33	0.32
Physical quantities per capita							
(a) Newspapers (copies per week)	2.8	–	–	–	–	2.9	3.0
(b) Newsprint (lbs per year)	59.0	–	–	–	–	12.0	13.0
(c) Magazine purchases (copies per month)	2.5	–	–	–	–	2.0	2.1

Source: Combined Production and Resources Board (1945), p.120.

a serious and absurd underestimate of the Print Order from Transport House for reading material. Every twopenny or threepenny pamphlet at open air meetings is immediately snapped up and usually paid for with silver, for which no change is asked. I believe if we had in stock 40,000 Penguins we could have sold them easily. What we sell is taken home and thrown back at us at the next open air meeting in the form of precise and thoughtful questions (quoted in Harrington and Young, 1978, p.66).

The third and final way in which changes in accounts came about, and the most difficult to pin down, was through the responsibility that men and women were forced to take for their lives. The pressures of this responsibility forced them to learn more about themselves and the world.

I don't expect many changes in <u>practical</u> form - but I do think the war has already shaken many people, once apathetic because they felt secure and did not know other people's lives, into at least an interest in securing healthier and less degrading conditions for the country (quoted in Calder and Sheridan 1984, p.210).

Some working class people lost the habit of 'deference' and again this must have had an effect on political accounts.

A Postscript; Fulham in 1945

In 1945 the two Fulham constituencies (East and West) were predominantly artisanal working class but with a substantial lower-middle class presence made up of shopkeepers, clerks and so on (Pelling 1967). The constituencies had a solid record of voting Conservative until 1938 when, in a by-election caused by death, Fulham West returned a Labour candidate. In the 1945 election both constituencies recorded a large swing to Labour and substantial Labour majorities (Table 14) What were the accounts being offered in the months up to the Election?

The accounts of many in the two constituencies, which had been much changed by the War and its damage, had not changed very much at all. Class solidarity was all.

Table 14

Fulham Election Results, 1935 and 1945

	Date	Electors	Turnout	Candidates	Votes Cast	%
Fulham East	1935	50,682	71.9	W.W. Astor (Con.)	18,743	51.4
				J.C. Wilmot (Lab.)	17,689	48.6
	1945	38,311	73.8	R.M.M. Stewart (Lab.)	15,662	55.4
				W.W. Astor (Con.)	10,309	36.4
				P.M. Synett (Lib.)	2,315	8.2
Fulham West	1935	49,480	69.9	C.S. Cobb (Con.)	18,461	53.4
				M. Follick (Lab.)	14,978	43.3
				W.J. Johnia (Lib.)	1,138	3.3
	1938	48,469	65.5	E. Summerskill (Lab.)	16,583	52.2
				C. Busby (Con.)	15,162	47.8
	1945	41,329	76.3	E. Summerskill (Lab.)	19,537	61.9
				P. Lucas (Con.)	12,016	38.1

Source: Craig (1969)

> Well, I think that its out to help the working
> class (Mass-Observation, File Report 2270A,
> p.107).
> Well, Labour's for the workers and its no
> good being a worker and voting for the
> bosses (Ibid., p.107).
> Well, its supposed to be for the working man
> and not against him (Ibid., p.107).
> Well, I'm a working man myself. I think a
> working man that votes Conservative is barmy
> (Ibid., p.107).

Most of these voters had always voted Labour:

> 'Well, I've always voted Labour. I've been
> a Labour man all my life'. (Ibid., p.111).
> 'I wouldn't think of voting any other way.
> My family would never forgive me'. (Ibid.,
> p.111).

The chief issue that had attracted previously
non-Labour voters as well as reinforcing the
accounts of previously Labour voters was housing,
of which there was a critical shortage in the constituencies.

> Oh, I'm voting for the Labour woman, and I'll
> tell you why. I think there'll be a proper
> bloody revolution when the boys come home,
> if something isn't done about the housing
> and by all I've read it's only Labour that's
> got a policy about the housing. (Ibid.,
> p.109).

> Housing, most of all. I was browned off of
> Churchill's last speech, nagging about the
> birth rate again. Give people houses and
> the jobs and they'll have the kids. Have you
> seen the boxes they're putting up for people
> to live in? (Ibid., pp. 109-110).

> I think all the parties will have to do some-
> thing about the rehousing - they'll have to
> make some attempt. The mood of the people
> won't be trifled with! (Ibid., p.109).

Nationalisation also sparked off some pro-
Labour accounts.

> Well, I'm going to vote Labour because I
> believe it's the only party likely to get

131

> things done and I believe that in the country's
> interest it would be best to nationalise the
> mines and public utilities. Speaking quite
> plainly, I think these measures are very much
> overdue. (Ibid., p.89).

Finally, the swing of the pendulum effect
was evident, especially amongst middle-class voters.

> From a business point of view I ought to vote
> Conservative, but somehow I feel that the
> Conservative party has had its run; we want
> something wider, and despite the fact that
> Labour doesn't represent the views of the
> people 100%, it is more likely to benefit
> the country as a whole. (Ibid., p.89).

These voters, who were previously non-Labour,
had spent considerable time talking about which
way to vote and in addition had read more. 'Well
I think I'm right in saying I've reached this
decision by simply reading and thinking about it'
(Ibid., p.89).

All in all, this section has tried to show
in the most preliminary way - in principle rather
than in practice - how contexts and accounts and
folk models changed in the circumstances of the
Second World War, leading to a particular political
outcome. Three final points are in order. First,
more work needs to be done before it is possible
to be certain how and in what ways people's accounts
and folk models did change. Second, it is vital
to reemphasise the extent of this change. It was
limited to a minority - a large minority - but
still a minority. Third, it is important to stress
that 'political' accounting is not some sacred
activity to be divorced from the secular flow of
other kinds of talk.

> We sat on the slope of the Head to watch the
> circus, and I saw a group sitting near in
> very earnest conversation, with their heads
> together. I'd have loved to go and butt in.
> I love being in an argument, and thought,
> 'Perhaps they are talking about the atomic
> bomb - or the result of the Election'. I've
> very good hearing, and when I'd got used to
> the different sounds around, I could hear
> what they were discussing - the new 'cold
> perm'! (Broad and Fleming (981, p.303).

LITTLE GAMES AND BIG STORIES:

Conclusion

A worked-out conception of the human agent is an
important part of any theory of social action
and this is something that has now increasingly
come to be realised in many areas of social science.
Geras (1983, p.107) states the case for historical
materialism but his strictures apply as well to
any other social theory:

> There are features of the (social) relations
> ...that are due precisely to the <u>nature of</u>
> <u>the entities they relate</u>, that is to say,
> to the general make up of human beings, to
> human nature. The latter is therefore a con-
> stitutive element in any concept of the
> ensemble of social relations, a view of it
> either explicit or implicity, absolutely
> necessary to any social theory, and discounting
> its theoretical role while simultaneously
> talking, under whatever name, about human
> society, a logical absurdity. The supposed
> replacement of the idea of human nature by
> the actual concepts of historical materialism
> ...is merely bombast. It is true certainly
> that one can overstate how much it is possible
> to explain by reference to human nature...
> it is also true that one can understate it
> and, in the context of historical materialism,
> the temptation to do so has become endemic.

In this paper I have tried to take note of these
strictures. I have outlined how the human agent
can be satisfactorily rendered as a creative discursive and social being and, in the subsequent
empirical example, I have considered a period of
history in which some persons were, more than is
usually the case, forced to face up to the world
with the result that their accounts and folk models
of themselves and the world were changed. The challenge is to extend this work to how these persons'
constructions of themselves as persons and selves
changed as the telling of these accounts and models
proceeded.

I want to conclude by making two more points,
both about class. First, it seems important, that,
if 'class consciousness' is to be located anywhere,
rather than being characterised as a generalised
amorphous sentiment emanating out of a population
like some kind of ectoplasm or as imagery distilled
from dead questionnaires(28), then it must be seen

as part of the discursive process of persons' accounts and folk models. This is, I believe, a conclusion quite similar to that reached by other writers recently (for example, Stedman Jones; 1983a, Joyce, 1984; Langton 1984)(29).

> Language disrupts any simple notion of the deformation of consciousness by social being because it is itself part .of social being. We cannot therefore decode political language to reach a primal and material expression of interest since it is the discursive structure of political language which conceives and defines interest in the first place. What we must therefore do is to study the production of interest, identification, grievance and aspiration within political languages themselves (Stedman Jones 1983a, pp.21-22).

This task is particularly crucial in English society because 'one of the peculiarities of England has been the pervasiveness of the employment of diverse forms of class vocabulary' (Stedman Jones, 1983a, p.2). Second, a discursive concept of class consciousness gives a more concrete expression to notions of community 'bonding' (Williams 1983). It shows that this bonding is created in a very complex and diffuse way which can only rarely be reduced to a set of simple solidary links. One of the problems with so many studies in recent years on the origins of, class consciousness has been that they have concentrated on a small tightly-knit set of contexts - from weavers to miners - where such links are readily and too easily apparent (Calhoun 1982). But the problem with modern societies is that they consist of sets of contexts which are more diffuse and less localised within which common elements of perception are developed in more diffuse and less localised ways. This problem of community is one that will be taken up in the next in this series of papers.

NOTES

1. For example, current political debates on the future of the left nearly all assume a fearsomely academic view of agents. These debates are therefore too often irrelevant.
2. As I hope to show, the definition of the 'political' is problematic in an account-based conception of the human agent.

3.   None of this means that a category like
'ideology' or 'hegemony' might not prove useful
but only given a clearly prescribed theory of social
action which as yet does not exist within the
marxist tradition.   There have of course been
marxist attempts to account for the individual
but these tend to veer between extreme reductionism
(e.g. Seve 1978) and extreme generality (e.g.
Leonard 1984).   A similar kind of reduction to
those imposed by categories like 'ideology' or
'hegemony' is that of classes having 'interests',
a viewpoint nicely demolished by Hindess (1982).
4.   A bloodline that starts with Vico and
works on through luminaries such as Heidegger and
the later Wittgenstein to sociologists like
Garfinkel, Mills and, latterly, Giddens.
5.   Further examples of power 'games' can
be found in Henriques et al (1984).
6.   The constructivist model is usually associated with the work of Harré and Shotter in
social psychology but elements of it can be found
in many other disciplines, for example in the work
of Garfinkel and Giddens in sociology, or Bourdieu
and Geertz in anthropology.
7.   That is, knowledge learnt from within
a social system, like that gained from walk-
ing the streets of a city, rather than knowledge
learnt from without, as from a city map (see Thrift
1985a).
8.   The intention is both to abolish the
distinction between an inner and outer characteristic of Cartesian thinking and lessen the emphasis
on an 'unconscious'.
9.   I have taken the terms 'a political economy of development opportunities' and 'a political
economy of selfhood' from Shotter (1984).
10.   See, for example, Elias (1978); Harré
(1979; 1983); Tuan (1983). I mark myself as myself
according to cultural norms.
11.   The crucial elements of these narrations
are the metaphor and the metonym (see Thrift 1985a).
12.   See, for example, the work on political
and other forms of socialisation such as that of
Stevens (1982) and Stacey (1978) and in the volumes
edited by Tajfel (1984) and Doise and Pulmonari
(1984).   It is in these institutions that story
telling begins to take on some 'theoretical' characteristics.   This is not a subject which I have
space to explore here but it is clearly important
(see Luria 1979; Ong 1982; Thrift 1985a).
13.   Quotes taken from Mass-Observation (1947,

1950).

14. 'Left' is a term I have purposely kept fuzzy in this paper for reasons that will become clear.

15. These accounts had already begun to change early on in the War. In 1942 for example, a Gallup Poll conducted entirely among the civilian population, already showed a marked left-wing lead (Pelling 1984).

16. In addition the Parliamentary Party could count on three ILP members of Parliament and one Commonwealth member of Parliament to vote with it.

17. Clearly geographical comparison of the 1935 and 1945 election results is complicated by the presence of the National Government in 1935. But the 1935 election is generally recognised as inaugurating the period of Labour-Conservative dominance in British politics. The National Labour Party won only 339,811 votes and gained only eight seats compared with the Labour Party's 8,325,491 votes and 154 seats (see Kinnear 1968).

18. For example, a Gallup Poll of June 1942 found that 62 per cent of those questioned thought that Russia was more popular with the British than the United States.

19. Unemployment had decreased from 1,270,000 in mid-1939 to 103,000 by mid-1945. Of course, some districts had already been doing well before the War (see Richardson 1967).

20. In terms of labour supply, regions were designated as 'importing', 'exporting' or 'self sufficient' in 1943 whilst local offices of the Ministry of Labour had been designated from 1941 as 'green', 'amber', 'red' or 'scarlet', according to the intensity of their mainly female labour requirements (Ministry of Labour & National Service 1947).

21. Short-term migration, like evacuation, was not well picked up in official statistics during the war.

22. There was a general increase in union militancy over the course of the War.

23. Although the Women's Institute retained its basic class character, the Red Cross was much less class divided because of the stringencies of examinations. For a number of interesting studies of voluntary organisations during the War, see Bottomore (1954); Ferguson and Fitzgerald (1954); Chambers (1959).

24. The official evacuees came disproportion-

ately from the poorest strata of society. Of course, some better off working-class people were billetted with poorer working-class people as well. This caused friction too.
   25. Few could afford them.
   26. This increase in readership should not be exaggerated. See Mass-Observation File Reports 47, 48, 227oc.
   27. This has been described as a political portent.
   28. The questionnaire tendency has been well criticised by Emmison (1985).
   29. However these writers go too far at times in making language the fount of all action.

ACKNOWLEDGEMENTS

I would like to thank the trustees of the Panty Fedwen Fund of Saint David's University College for providing me with funds to enable the diagrams to be drawn and to visit the Mass-Observation Archive at the University of Sussex. I also acknowledge the generous help given by the Archive itself. Robin Butlin, Denis Cosgrove, Derek Gregory, Ron Johnston, Malcolm Smith and John Shotter provided sources, encouragement and understanding.

REFERENCES

Addison, P. (1975) The Road to 1945. British Politics and the Second World War, Jonathan Cape, London.
Anderson, B. (1983) Imagined Communities. Reflections on the Origin and Spread of Nationalism, Verso, London.
Bernstein, R.J. (1983) Beyond Objectivism and Relativism. Science, Hermeneutics and Praxis, Blackwell, Oxford.
Billinge, M. (1984) 'Hegemony, class and power in late Georgian and early Victorian England: towards a cultural geography'. In A.R. Baker and D. Gregory (eds.), Explorations in Historical Geography. Interpretative Essays, Cambridge University Press, Cambridge, pp. 28-67.
Bonham, J. (1954) The Middle Class Vote, Faber and Faber, London.
Bottomore, T. (1954) Social stratification in voluntary organisations. In D.V. Glass (ed.), Social Mobility in Britain, Routledge and

Kegan Paul, London, pp. 349-382.

Bourdieu, P. (1977) Outline of a Theory of Practice, Cambridge University Press, Cambridge.

Broad, R. and Fleming, S. (1981) Nella Last's War. A Mother's Diary 1939-45, Sphere, London.

Butler, D. and Stokes, D. (1969) Political Change in Britain, Macmillan, London.

Calder, A. (1969) The People's War. Britain 1939-1945, Jonathan Cape, London.

Calder, A. and Sheridan, D. (1984) Speak for Yourself. A Mass-Observation Anthology 1937-49, Jonathan Cape, London.

Calhoun, C. (1982) The Question of Class Struggle, Blackwell, Oxford.

Calvocoressi, P. (1978) The British Experience 1945-75, Bodley Head, London.

Central Statistical Office (1951) Statistical Digest of the War, HMSO, London.

Castoriadis, C. (1984) Crossroads in the Labyrinth. Harvester, Brighton.

Chambers, R.C. (1954) 'A study of three voluntary organisations'. In D.V. Glass (ed.), Social Mobility in Britain, Routledge and Kegan Paul, London, pp. 353-466.

Childs, D. (1979) Britain since 1945. A Political History, Benn, London.

Cohen, J.L. (1903) Class and Civil Society. The Limits of Marxian Critical Theory, Martin Robertson, Oxford.

Combined Production and Resources Board (1945) The Impact of the War on Civilian Consumption, HMSO, London.

Connell, R.W. (1983) Which Way is Up? Essays on Class, Sex, and Culture, George Allen and Unwin, Sydney.

Craig, F.W.S. (1969) British Parliamentary Results 1918-1949, Parliamentary Research Service, Chichester.

von Cranach, M. and Harré, R. (eds.) (1982) The Analysis of Action: Recent Theoretical and Empirical Advances, Cambridge University Press, Cambridge.

Doise, W. and Palmonari, A. (eds.) (1984) Social Interaction in Individual Development, Cambridge University Press, Cambridge.

Eatwell, R. (1979) The 1945-1951 Labour Governments, Batsford, London.

Elias, N. (1978) The Civilising Process, Blackwell Oxford.

Emmison, M. (1985) 'Class images of the economy: opposition and ideological incorporation within working class consciousness', Sociology, 19, pp. 19-38.

Ferguson, S. and Fitzgerald, H. (1954) Studies in the Social Sciences, HMSO, London.

Fogarty, M.P. (1945) Prospects of the Industrial Areas of Great Britain, Methuen, London.

Gallup, G.H. (1976) The Gallup International Public Opinion Polls. Great Britain 1937-1975. Volume One 1937-1964, Random House, New York.

Gauld, A. and Shotter, J. (1977) Human Action and its Psychological Investigation, Routledge and Kegan Paul, London.

Geertz, C. (1983) Local Knowledge. Further Essays in Interpretative Anthropology, Basic Books, New York.

Geras, N. (1983) Marx and Human Nature. Refutation of a Legend, Verso, London.

Giddens, A. (1979) Central Problems in Social Theory. Action, Structure and Contradiction in Social Analysis, Macmillan, London.

Giddens, A. (1982) 'Commentary on the debate', Theory and Society. 11, pp. 527-539.

Giddens, A, (1984) The Constitution of Society, Polity, Cambridge.

Gregory, D. (1984a) 'Contours of Crisis? Sketches for a geography of class struggle in the early Industrial Revolution in England'. In A.R. Baker and D. Gregory (eds.), Explanations in Historical Geography. Interpretative Essays, Cambridge University Press, Cambridge, pp. 68-117.

Gregory, D. (1984b) People, places and practices: the future of human geography (mimeo).

Hall, S. (1983) 'The problem of ideology' in E. Matthews (ed.), Marx: A hundred years on. Lawrence and Wishart, London, pp. 50-63.

Harré, R. (1974) 'Blueprint for a new science', in N. Armistead (ed.) Reconstructing Social Psychology, Penguin, Harmondsworth, Middlesex, pp. 240-259.

Harré, R. (1979) Social Being. A Theory for Social Psychology, Blackwell, Oxford.

Harré, R. (1983) Personal Being. A Theory for Individual Psychology, Blackwell, Oxford.

Harrington, W. and Young, P. (1978) The 1945 Revolution, Davis Poynter, London.

Harrison, T. (1942) 'Class consciousness and class unconsciousness', Sociological Review, 34.

Henriques, J., Holloway, W., Urwin, J., Couze, R., Walkerdine, V. (1984) Changing the Subject, Methuen, London.

Heritage, J. (1984) Garfinkel and Ethnomethodology, Polity, Cambridge.

Hindess, B. (1982) 'Power, interests and the outcome of struggles', Sociology, 16, pp. 498-511.

Hoban, R. (1983) Pilgermann, Jonathan Cape, London.

Holy, L. and Stuchlik, M. (eds.), (1981) The Structure of Folk Models, Academic Press, London.

Holy, L. and Stuchlik, M. (1983) Actions, Norms and Representations. Foundations of Anthropological Inquiry, Cambridge University Press, Cambridge.

Howard, A. (1963) 'We are the masters now' in M. Sissons and P. French (eds.), Age of Austerity 1945-1951, Hodder and Stoughton, London.

Joyce, P. (1984) 'Languages of reciprocity and conflict: a response to Richard Price', Social History, 9, pp. 225-231.

Kinnear, M. (1968) The British Voter. An Atlas and Survey since 1885, Cornell University Press, Ithaca, New York.

Kirby, A. (1985) 'Pseudo-random thoughts on space, scale and ideology in political geography'. Political Geography Quarterly, 4, pp.5-18.

Langton, J. (1984) 'The industrial revolution and the regional geography of England', Transactions, Institute of British Geographers, 9, pp. 145-167.

Lash, S. and Urry, J. (1984) 'The new Marxism of collective action: a critical analysis', Sociology, 18, pp.33-50.

Leonard, P. (1984) Personality and Ideology. Towards a Materialist Understanding of the Individual, Macmillan, London.

Luria, A.R. (1979) The Making of Mind, Harvard University Press, Cambridge, Mass.

McCallum, R.B. and Readman, A. (1947) The British General Election of 1945, Oxford University Press, London.

Madell, G. (1981) The Identity of the Self, Edinburgh University Press, Edinburgh.

Marwick, A. (1982) British Society since 1945, Pelican, Harmondsworth, Middlesex.

Mass-Observation (1939) Britain, Penguin, Harmondsworth, Middlesex.

Mass-Observation (1940) War Begins at Home, Chatto and Windus, London.

Mass-Observation (1942) People in Production, Advertising Service Guild, London.

Mass-Observation (1943a) People's Homes, Advertising Service Guild, London.

Mass-Observation (1943b) War Factory, Gollancz,

London.

Mass-Observation (1947) Puzzled People. A Study in Popular Attitudes to Religion, Ethics, Progress and Politics, Gollancz, London.

Mass-Observation (1950) The Voter's Choice. Art and Technics, London.

Merchant, C. (1980) The Death of Nature: Women, Ecology and the Scientific Revolution, Harper and Row, New York..

Mills, C.W. (1940) 'Situated actions and vocabularies of motive', American Sociological Review, 5, pp. 904-913.

Ministry of Labour and National Service (1947) Report for the Years 1939-1946. HMSO, London, Command 7225.

Minns, R. (1980) Bombers and Mash. The Domestic Front 1939-45, Virago, London.

Nagel, T. (1979) Mortal Questions, Cambridge University Press, Cambridge.

Ong, W.J. (1982) Orality and Literacy. The Technologising of the Word, Methuen, London.

Padley, R. and Cole, M. (1940) Evacuation Survey. A Report to the Fabian Society, George Routledge, London.

Parfit, D. (1984) Reasons and Persons, Clarendon Press, Oxford.

Pelling, H. (1967) Social Geography of British Elections, Macmillan, London.

Pelling, H. (1984) The Labour Governments 1945-51, Macmillan, London.

Philo, C. (1984) 'Reflections on Gunnar Olsson's contribution to the discourse on contemporary human geography', Society and Space, 2, pp. 119-248.

Pimlott, B. (1985) 'The road from 1945', The Guardian, 27 July, 15.

Planning (1948) 'Manpower movements' Planning, 14, Pamphlet 276.

Prigogine, I. (1980) From Being to Becoming. Time and Complexity in the Physical Sciences, W.H. Freeman, San Francisco, California.

Prigogine, I. and Stengers, I. (1984) Order out of Chaos, Bantam Books, New York.

Reiss, D. (1981) The Family's Construction of Reality, Harvard University Press, Cambridge, Mass.

Richardson, H.W. (1967) Economic Recovery in Britain, 1932-9 Methuen, London.

Sayer, A. (1984) Method in Social Science. A Realist Approach, Hutchinson, London.

Secord, P.F. (ed.) (1982) Explaining Human Behaviour. Consciousness, Human Action and Social Structure. Sage, Beverly Hills.

Seve, L. (1978) Man in Marxist Theory and the Psychology of Personality, Harvester, Brighton.

Shotter, J. (1984) Social Accountability and Selfhood, Blackwell, Oxford.

Shotter, J. (1985a) 'Speaking practically: a contextualist account of psychology's context' in Rosnow, R. and Georgodi, M. (Eds.), Contextualism and Understanding, Praeger, New York.

Shotter, J. (1985b) 'Accounting for place and space', Environment and Planning D, Society and Space, 3, pp. 447-460.

Stedman Jones, G. (1983a) Languages of Class, Studies in English Working Class History 1832-1982, Cambridge University Press, Cambridge.

Stedman, Jones, G. (1983b) 'Rethinking Chartism' in Stedman Jones, G. Languages of Class. Studies in English Working Class History. 1832-1982. Cambridge University Press, Cambridge, pp. 90-178.

Stevens, O. (1982) Children Talking Politics, Martin Robertson, Oxford.

Stevenson, J. (1984) British Society 1914-45, Pelican, Harmondsworth, Middlesex.

Swift, G. (1983) Waterland, Heinemann, London.

Tajfel, H. (ed.), (1984) The Social Dimension, 2 vols. Cambridge University Press, Cambridge.

Thrift, N.J. (1979) 'Limits to knowledge in social theory: towards a theory of human practice', Australian National University, Department of Human Geography, Seminar Paper.

Thrift, N.J. (1983a) 'Literature, the production of culture and the politics of place'. Antipode, 15, pp. 12-24.

Thrift, N.J. (1983b) 'On the determination of social action in space and time', Environment and Planning D. Society and Space, 1, pp.23-57.

Thrift, N.J. (1983c) 'Editorial. The politics of context', Environment and Planning D. Society and Space, 1, pp.371-376.

Thrift, N.J. (1985a) 'Flies and germs. The geography of knowledge'. In D. Gregory and J. Urry (eds.), Social Relations and Spatial Structures, Macmillan, London, pp.366-403.

Thrift, N.J. (1985b) 'Bear and mouse or bear and tree, Anthony Giddens' reconstitution of social theory', Sociology, 19, pp. 609-623.

Thrift, N.J. (1987) Social Theory and Human Geography, Polity, Cambridge.

Titmuss, R. (1950) Problems of Social Policy, HMSO, London.

Tuan, Yi-Fu (1983) Segmented Worlds and Self, University of Minneapolis Press, Minneapolis.

Vygotsky, L.S. (1962) Thought and Language, MIT Press, Cambridge, Mass.

Vygotsky, L.S. (1978) Mind in Society. The Development of Higher Psychological Processes, Harvard University Press, Cambridge, Mass.

Williams, B. (1973) Problems of the Self, Cambridge University Press, Cambridge.

Williams, R. (1977) Marxism and Literature, Oxford University Press, Oxford.

Williams, R. (1983) Towards 2000, Chatto and Windus, London.

Wittgenstein, L. (1965) The Blue and Brown Books, Harper and Row, New York.

Wittgenstein, L. (1980) Remarks on the Philosophy of Psychology, Volume 1. Blackwell, Oxford.

CHAPTER SIX

STATE SPONSORED CONTROL - MANAGERS, POVERTY
PROFESSIONALS AND THE INNER CITY WORKING CLASS

DAVID BYRNE

Introduction

'The long-term aim of the partnership for the inner
area is to secure an improved economy and to make
it a place where people want to live and work.
In the interim, the partnership will act to al-
leviate the worst aspects of the disadvantages
currently experienced by inner area residents'.
(Newcastle/Gateshead - Inner City Partnership -
Action Programme 1984-1987, p.2).
    This paper is an attempt to comprehend the
relationship between the process of the restructuring of the working class in a metropolitan capitalist county consequent upon the reorganisation
of capitalist production on a world scale (in short-
hand de-industrialisation) on the one hand, and
the form and content of its administration on the
other. The argument advanced is that with the decline of 'work' as a central experience for an
ever increasing marginalised 'social proletariat'
the regulation of life becomes more and more a
function of administrative decisions. The development and operation of these mechanisms is in contradiction to democratic principle and practices.
Essentially they are 'capitalist'. However, whereas
the general pattern of the development of corporatism has involved the cooption of working class
leadership in a 'representative but non-accountable'
capacity, de-industrialisation and class disorganisation has meant that such cooption of the representatives of the 'organised' or 'central' working
class is not enough. Capitalism and the capitalist
state have to have a relationship with the 'disorganised' and 'peripheral'. However, the very
restructuring which corporate capital has imposed
on the working class as a mechanism for resolving

144

the nature of the current crisis means that such cooption is difficult in the extreme.

Something else that needs to be said at this stage is that what has been written this far is not innocent. A number of 'positions' on issues of analysis have been asserted without specific identification or reference to extensive intellectual and political controversy. This is quite deliberate. My own view is that the major problem with debates in this area in general, and with the debate about the changing nature of the state as a body of practices within capitalism in particular, is that they are far too little informed by empirical example. It is in the nature of the work in which I am engaged that I have to come at things from the opposite direction. I see what is happening and have to try to make sense of it. To my mind this has decided advantages.

This piece is therefore empirical. It is a kind of reflection upon the things that are going on under the aegis of the Newcastle-Gateshead Inner City partnership in an attempt to set out what is happening, and what is to be done about it. I will proceed to describe and conclude by reflecting.

In 1981 the two Metropolitan Districts of Newcastle and Gateshead had a combined population of approximately 484,000. Of these 198,000 lived in the 'Partnership Area'. The population of the 'Partnership Area' has been declining since the 1960s when most of the 'Districts' population lived in these areas. Indeed since 1971 the 'Partnership Area' has lost nearly 20 per cent of its population. The population of the 'rest' of the Districts has remained broadly static. Table 15 gives some general information about 'Inner' and 'Outer' Newcastle/Gateshead.

What has become necessary is a reorganisation of the working class so that a relationship can be established with the marginalised 'social proletariat', because such a relationship is vital to the continuing maintenance of the administrative structures of the state which are concerned with facilitating the reproduction of the 'social proletariat' within capitalist relations of production. The way in which an attempt is being made at the resolution of this problem in the urban United Kingdom is through the development of corporate management practices in local administration. These are associated with the 'social proletariat' through the operations of 'poverty professionals', i.e.

145

Table 15:

COMPARING BETWEEN 'INNER' AND 'OUTER

	Partnership Area	Rest of Newcastle/ Gateshead
Tenure of households		
% owner occupied	21.4	51.1
% L.A. tenants	60.2	37.1
% H. Association	4.4	2.2
% Other tenants	14.0	9.6.
Housing amenities		
% Lacking 1 or more of 3 basic	2.1	1.0
% Persons living more than one per room	7.1	3.0
Age structure		
% Pensioners	25.7	24.2
% Population 0-15	22.4	22.8
% Social class by household head		
i	1.9	6.8
ii	11.0	24.6
iii N.M.	10.7	15.1
iii M.	29.0	33.6
iv	21.1	13.5
v	12.4	5.0
Other	3.9	1.4
No car %	66.6	39.6
'Industrial' Workers	67.1	59.5
% Economically active - males		
Working	74.5	87.3
Seeking work	23.0	11.4
Temporarily sick	2.5	1.3
% Economically active - females		
Working	88.6	93.5
Seeking work	9.9	5.7
Temporarily sick	1.4.	0.8

Source: Newcastle/Gateshead Inner City Partnership
(all data from 1981 census)

state employees (whether directly or indirectly
employed), whose task it is to create organisation
where none exists, in order to facilitate the administration of the disorganised. To put this in
plain English, community work is about putting
together what capitalism has broken up, but on
capitalism's terms. This is a contradictory process.

In writing this in 1985 I am very conscious
of to what very considerable extent what has just
been written is a restatement of Cynthia Cockburn's
argument in The Local State (1977). There are however, two new aspects which are worth highlighting.
The first is the explicit use of the term and concept 'social proletariat'. The idea was in fact
implicit in Cockburn's argument but its explicit
assertion is important. The second is that the
extent of de-industrialisation and disorganisation
is now far greater than it was in the middle 1970s.
Trends which Cockburn and others (notably the Home
Office Community Development Projects) identified
have now developed to such an extent that the situation has been qualitatively transformed.

The first thing to realise is that while contrasts were there, they were not that strong in
1981. Basically this was a function of the boundary definitions of the inner city Partnership Area
and of the general de-industrialisation of the
conurbation economy. The appropriate contrast
for Bensham in Gateshead is not Leam Lane in Felling
but Surbiton. By September 1983 male unemployment
in the Partnership Area stood at 25.5 per cent
(Department of Employment definition) compared
with 17.0 per cent in the rest of Gateshead (Gateshead Director of Planning 1983). The point of all
this data is simple enough. It is to illustrate
that by the beginning of the 1980s inner city
Gateshead and Newcastle was a place in which Claus
Offe's prophecy was becoming increasingly true.

> ...in response to your question concerning
> the potential of the labour movement, I think
> its potential has been exhausted to the extent
> that it ignores the fact that the wage-labour
> capital relationship is not the key determinant
> of social existence, and that the survival
> of capitalism has been increasingly contingent
> upon non-capitalist forms of power and con-
> flict... In the present period this problem
> has been made more acute because the vision
> of full employment has been undermined by

capital itself. (Offe 1984, p. 285).

I have to say here, very forcefully indeed, that I would insert the word immediate between 'key' and 'determinant' in this passage. I do believe, very firmly, that the wage labour-capital relationship is the determinant of social existence. Where I agree with Offe is in relation to the fact that, for more and more people, it is not experienced as such, because they are not employed as wage labour under capital even if all the conditions of their lives are (in Raymond Williams's, 1980, sense of the word) determined by the relationship between wage labour and capital. In other words, I am agreeing with Offe's account of immediate experience and profoundly disagreeing with his analysis of its fundamental implications.

The implication of this is very important:

> My thesis is that under modern capitalist conditions there is a no one central condition that causally determines all other conditions on a base-superstructure or primary - secondary manner. The work role is only partly determinative of social existence. (Offe 1984, p.283).

This paper is about the way in which 'control' is exercised when the 'discipline of labour' is not the key disciplinary experience of most people. It is also about a situation in which at a specifically local level: 'behind the facade of parliamentary democracy both political conflict and the resolution of policy issues increasingly take place within organisational settings which are unknown to democratic theory' (Offe 1984, p. 167).

Despite the prevalence of quotations from Offe in this introduction, this is not an 'Offian' piece. In crucial respects, as indicated above, the analysis presented differs from Offe's, but Offe is a clear and honest writer who can be used to define the issue from one direction. He talks about 'government' in the welfare state and that is what I am trying to deal with - the form and content of government with regards to both decision-making and day to day control. I am writing about how people are managed in an increasingly prevalent context.

Thus, this essay is not about the present character of 'economic initiatives' for the inner cities of the UK. For that see Anderson (1983)

and Shutt (1984). In a sense it is not about accumulation at all, although it is of course located
in a context circumscribed by the contemporary
logic of capital accumulation(1).. Rather it is
about 'reproduction' of a social order in relation
to crisis. It focuses on the relationship between
decision-making by 'urban managers' operating in
a corporatist environment and 'control' by a newish
group of state and quasi-state workers operating
in direct relation to the working class in locales
of the sort represented by 'Partnership' Gateshead
and Newcastle. In attempting the task I have set,
I will draw to a considerable degree on two themes
present in earlier work carried out by myself and
Don Parson. These themes are 'underdevelopment'
and 'substitution'. Thus this paper is about the
nature of local 'politics' in an accumulation context which can be characterised as one of 'underdevelopment' and an administrative context of 'substitution'. What do these terms mean on Tyneside
in 1985?

'Underdevelopment and Substitution', or 'Where Did the Power Go?

Don Parson and myself have elaborated the notion
of 'underdevelopment' (which we derive directly
from the work of Harry Cleaver 1977) in a recently
published paper (Byrne and Parson 1983). That paper
was concerned with 'change in the form of the working class which are being forced on that class
by the contemporary characters of capitalist production' (Byrne and Parson 1983, p. 127). It
attempted to relate the peripheralisation of production to the 'marginalisation' and emmiseration
of previously 'central' workers (see A. Friedman,
1977). What we generated was a particular version
of Cleaver's discussion:

> Development and underdevelopment are understood
> here neither as the outcome of historical
> processes (as bourgeois economists recount)
> nor as the processes themselves (as many Marx-
> ists use the terms). They are rather two
> different strategies by which capital seeks
> to control the working class...they are always
> co-existent because hierarchy is the key to
> capital's control and development is always
> accompanied by relative or absolute under-
> development in order to maintain that hier-
> archy... By development I mean a strategy

149

> in which working class income is raised in
> exchange for more work... The alternative
> strategy, in which income is reduced in order
> to improve the availability for work, I call
> a strategy for underdevelopment (Cleaver 1977,
> p. 94).

We therefore interpreted the 'de-industrialisation'
of U.K. in general, and of their inner cities in
particular, as part of a process of underdevelopment. In other words, the logic of capital was
forcing an even greater proportion of the working
class out of centrality and over the sharp ultra-
class divide into marginality. My favourite image
of this is a falling off a cliff - easy to do,
hard to get back up. The fact is that these people
are not falling: they are being pushed.

It is very important to understand that the
consequences of this are not just to be found at
the point of production. Indeed they are not just
to be located in production and reproduction (in
the simple sense of the domestic sphere). They
pervade the characters of politics and administration and underlie the development of corporatism.
In other words, the idea of 'underdevelopment'
provides an explanation of why the working class
have allowed these things to happen.

That action-centred phraseology is very deliberate. The whole basis of the 'autonomist perspective' is that the working class is the source
of crises in capitalism. Cleaver's 'political'
reading of Capital (1979) is essentially an assertion that the working class has to be understood
as being '...within capital, yet capable of
autonomous power to disrupt the accumulation process
and thus break out of capital' (Cleaver 1979, p.62),
coupled with a history of the capitalist mode of
production written in terms of 'the power relations
between classes'. It may seem strange to assert
'autonomy' in a paper which is a document of 'de-
feats' but, as O'Connor says, it is a matter of
'...what has been made to happen in history and
what can be made to happen in the future' (1981,
p.237).

Which leads us to the questions of what has
happened and why? This brings us to the concept
of 'substitution', which I want to locate in
relation to the development of corporate management
in local government. In summary, I would argue,
following Cockburn (1977; see also Bennington 1974;
Byrne 1982), that we can understand corporate

150

management as:

> ...a historically specific response generalised
> spatially by the Baines Report and the 1974
> reorganisation of local government, on this
> part of the U.K. national state which had
> the objective of coping with the problems
> posed by reformist labourist control of the
> local government of working-class areas (Byrne,
> 1982, p. 75).

Thus, 'substitution' means the replacement of democratic political control by corporate management and initiatives like partnership, which are things 'unknown to democratic theory'. I originally used this term (1982, p. 67) to refer to a process of de-localisation and although de-localisation is an aspect of what I am getting at, it is a means to 'de-democratisation' rather than the essential element of substitution.

Let me draw this together before turning to historical accounts. What I have been trying to do in this paper is: (1) to identify the nature of 'Partnership' Gateshead/Newcastle in terms of symptomatic description; and (2) to specify the content of two conceptual tools - 'underdevelopment' and 'substitution' - which I intend to employ in clarifying the nature of this locale, the reasons why it is as it is in 1984, and (much more importantly) why the people who live in it are being treated as they are and are putting up with that treatment. I want now to do two things. The first is to give an account of the emergence of 'Partnership' and of the character of the arrangements underlying it. Associated with this I want to describe the growth of, and character of, 'poverty professionalism'; of a range of new occupations whose task it is to mediate between the peripheralised working class and those who, formally, manage cities and people. Of course the poverty professionals are part of the management, a crucial part. They control.

Before getting down to history let me say why I chose to write this paper for a conference on space and stratification. We are going to have to get rid of the stratification image for class structure. The character of class is, within limits (which are transcendable), dynamic. It is changing. In 1961 when I was fourteen the dominant experience of Tyneside was full employment in an industrial economy. My fourteen year-old daughter lives in

a different world. A good bit of my world sur-
vives(2) but there is another world which is the
consequence of all those people being pushed off
that cliff, and that world is spatially located.
Look back at the tenure figures in Table 1. In
the 'Partnership' 80 per cent of households are
tenants. In 'the rest' 49 per cent are, and that
figure is falling quite rapidly because a great
bulk of council house sales since 1980 have been
ex-Partnership. There is a spatial concentration
of the dispossessed. Not all 'Partnership' is like
this, but the existence of locales like St.
Cuthbert's Village (with an unemployment rate for
men 38.0 per cent in September 1983), Saltmeadows
(32.0 per cent) or Old Ford (29.8 per cent), pose
management problems, especially when the mediating
commodity-based mechanisms of provision of the
means of reproduction have almost entirely dis-
appeared. It is the state which faces the people
in almost every aspect of their everyday lives.
The point is that their class location and spatial
location are inter-related. Let us look at the
state's response.

Partnership - or We'll Make You an Offer You Can't Refuse

Inner City 'Partnership' was an idea introduced
by Labour in 1977, but which has been quite en-
thusiastically continued by the Thatcher Government.
Its structure has now been simplified in response
to a 1981 review by 'three representatives of local
private sector interests' (Partnership Annual Report
1982). Basically there is a 'Partnership Steering
Group' including both councillors and officers
of the local authorities, civil servants, and
officers of various quangos (especially the
health service) which is, in effect, the Partnership Executive. The largest element of Partnership resources is devoted to 'economic development'. For the financial year 1984-1985 the
'Partnership' will spend £23.4 million, of which
£10.6 million (45%) is to be spent on 'economic
development', in a way which very well supports
Shutt's (1984, p. 38) assertion that:

> Compared then, to the rhetoric enterprise
> zones (and related Partnership activities)
> are heavily underwritten by the state and
> the evidence suggests that they are not per-
> forming functions which differ significantly

from traditional regional policies. They
are based on the assumption that 'bribing
capital' to shift locations and unloading
development costs for construction and finance
capital can actually halt the process of industrial restructuring which is creating chaos
in our conurbations and contribute to new
job creation.

Of the remaining £12.8 million, £8 million is being
spent on housing and environment, and community
services capital schemes. The remaining £4.8
million is spent on revenue schemes in these two
areas. It is on this that I want to focus, but
the context of this expenditure has to be located.
     Panitch (1980, p. 173) has defined 'corpor-
ation' as:

> ...a political structure within advanced capi-
> talism which integrates organised socio-econ-
> omic groups through a system of representative
> and cooperative mutual interactions at the
> leadership level and mobilisation and social
> control at the mass level. Seen in this way
> corporatism is understood as an actual politi-
> cal structure, not an ideology.

Offe (1984, p. 167) expands on this theme
in a way which is pertinent to our present discussion:

> Very often, decisions on key political issues
> emerge instead out of a highly informal process
> of negotiations among representatives of stra-
> tegic groups within the public and private
> sectors. Consultation, negotiation, mutual
> information and inconspicuous techniques of
> estimating potential resistance and support
> for a specific policy assume a role in public
> policy-making which is by no means restricted
> to a supplementary one. Apart from its highly
> informal character two other aspects are
> characteristic of this mode of public policy-
> making. One is the strong element of func-
> tional representation, and the other is its
> lack of democratic legitimation. Such para-
> parliamentary, as well as para-bureaucratic,
> forms of decision-making have therefore been
> described as neo-corporatist methods of
> interest intermediation. Corresponding to
> these characteristics, there is every reason

153

for the participants to keep their delicate exchange of proposals, information and threats as remote as possible both from the general public eye and from the segmental constituencies which participants represent.

Clearly the Partnership activities in support of capital accumulation are quite traditional and the structure corresponds well to the requirements of such sorts of policy formation and implementation. The involvement of local democratically elected representatives, or rather an elite of such representatives, is in itself of great importance. David G. Green (1981), in his examination of the policies of local government in Newcastle in the immediately pre-Partnership period, pointed out the anti-democratic character of proposals emerging from the Baines Report in a way reminiscent of both Bennington (1975) and Cockburn (1977), but as he was writing from a non-marxist background his approach highlights its affront to classic democratic theory and procedures. We marxists expect and get the worse but an old-fashioned liberal like Green gets shocked.

My own feelings about this emerge from historical work. There is no doubt whatever that the high period for democratic politics in Tyneside local government was from 1890 to 1939, or perhaps really 1918/1926 to 1939. In this period, a mass franchise operated in relation to real political issues dealing with the course of development (see Byrne 1979; North Tyneside CDP, 1977). Things are different now. Until one reaches the 1960s, an examination of Tyneside local government is an examination of change and development; however contradictory, partial and reformist those changes and developments were (I am here using development to mean change for the better). Since 1975 (the date that the IMF bulleted any serious reformist effects by the then Labour Government), the activities of the local state have been largely a matter of crisis management in the most general sense.

The point is that in such contexts a key aspect of legitimation disappears. An important part of the legitimacy of the democratic process has always derived from the fact that in some respects the democratic process has worked at the level of the local state. Voting left produced some material and cultural goodies. Since 1975 the inner city working class in Gateshead-Newcastle have voted even more solidly for Labour and it can be argued that the end result of that process, unlike in previous periods, has been no cultural or material gains whatsoever. Why then the awful quiet?

I am not suggesting that what I want to go on to discuss is the only explanation for the awful quiet, but it is one part of the explanation and it merits some attention.

Poverty - Professionalism - If You Can't Incorporate Them Control Them

In almost all of the Tyneside towns in the 1930s there were branches of the National Unemployed Workers Movement. The leading lights in these branches were a mixture of members of the Communist Party and Independent Labour Party, who believed in a socialist future, fought cases, campaigns and riots as occasion demanded, and subsisted personally and as an organisation on dole-money, collection and one penny subscriptions. They were against capitalism and for the future. When they controlled the local state (as with the Gateshead Poor Law Guardians in the 1920s), they used it as a weapon in battles; battles which were usually defensive and reformist but which were nonetheless battles.

Here is a list of 'new' projects for the financial year 1984-85 in the Gateshead-Newcastle Inner City Partnership, which are described under the heading 'community support':

1. John Marley Project (£80,000 capital + £20,000 revenue) - using a former comprehensive school in Scotswood (remember Vickers-Scotswood) as an adult education centre, to support locally controlled enterprises, to increase recreation and leisure facilities and to 'encourage participation by inner city residents in the development of their initiatives'.
2. Gateshead Voluntary Organisation Council - volunteer training programme (£14,000 revenue).
3. Gateshead Community Bus (£15,000 capital + £19,000 revenue) - for information and advice services.
4. Specialist Welfare Benefit Officers (£7,000 revenue) - for Disablement Information and Fuel Benefits.
5. School Attendance Social Work Team (£30,000 revenue) - that's an old idea!

6.   Kenton Community Police   Station   (£16,000 revenue).
7.   Single   homeless   <u>emergency</u>   accommodation (£141,000 capital).

Interesting ongoing schemes include: two law centres which are block funded from priority area teams, eight or ten priority estate action teams (it might be as many as thirteen depending on how you count); ethnic minorities projects; community centres, etc.   In other words, there are a massive number of small projects based on resources distributed to 'interest groups' and some rather larger 'management' exercises operating in this small geographic area.   Under the heading 'community support' alone there are some 36 extant projects. The question is what is the overall effect of all this going on.   It is worthwhile taking a kind of mental rest break here.   I am not suggesting that this sort of system is the consequence of a right wing plot.   Indeed, although seldom funded under Partnership arrangements (which include the provision, 'Finally, and very exceptionally the government may decide to withdraw support if a scheme becomes involved in a partisan political activity, since that could bring the Urban Programme as a whole into dispute', Partnership, July, 1981:3), there have been left 'client grants', where arrangements with (e.g. the GLC)   are not unlike those of 'un-political' groups with Partnership.   There are disturbing similarities between 'neighbourhood management' (see Seabrook 1984) and intensive housing management.   I am, I think, dealing with a general tendency rather than a specific conspiracy.   So, back to the question.
A lot of people, many of them 'local', or 'grass-roots', or 'hired activists' are employed on projects of the sort listed below.   What are they doing?   I can best attempt an answer to that question by telling a story.   In the middle 1970s North Tyneside CDP (one of the somewhat less than twelve reasons why the 'disrepute' clause is written into Partnership) was working with a tenants' group on the South Meadowell Estate in North Shields (see N. Tyneside CDP 1977, Vol. 3). Community work was simple minded then.   The tenants thought they were having to live in rotten houses (or rather flats).   The team agreed and provided the tenants with some organisational and research support. The tenants confronted the local (Labour-controlled) authority and they won!!   This was an example of

poor people's movements and how they succeed (see
Piven and Cloward 1979).
It is now 1984. There is an estate in
Gateshead called St. Cuthbert's Village, which
is to how the South Meadowell was in 1975 as hell
is to purgatory. In the past there had been tenants
groups on this estate directed not at its improve-
ment but at its eradication. Now there is a 'comprehensive management team' costing £95,000 a year
and dealing with voids, difficulties and deviants.
These sorts of things have been written about elsewhere (see Byrne and Parson 1983). The point is
that whereas the style of community work in the
middle 1970s was political, now, in an authority
where 58 out of 66 elected representatives are
Labour, it is managerial. At least this is reasonably obvious.
Consider instead community law centres. One
of the two in the Partnership area is actually
part of the residue of Benwell CDP of the 1970s.
These are fully Partnership funded - (in
Gateshead's case as a consequence of local initiative) - and are locally managed by committees which
include a minority of local authority councillors.
The law centres take up people's time. They exist,
they advocate, they argue for the poor, but in
the end '...legal decisions are made in a political
and economic context. The issues are essentially
political ones. It is the reality behind the law
which must be challenged' (North Tyneside CDP 1977,
p. 4).
Of particular interest in this context is
Newcastle's 'priority areas programme' which pre-
dates the establishment of the Inner City Partnership, but is closely associated with it in practice.
This has been fully described by Chris Miller (1981)
and I am drawing on that description here. Basically
the priority area programme (also known as 'stress
areas project') is an area management scheme involving the extension of the corporate approach
to an area level. It originated as one of the
Department of Environment's area management trials,
but as Miller (1981, p. 97) comments:

> Newcastle's approach...is markedly different
> from many of the other schemes, being estab-
> lished explicitly to tackle deprivation
> (labelled 'stress' by the city) within the
> inner area. Indeed the authority had already
> rejected an area management project styled
> after the Liverpool 8 scheme. The key feature

of the programme has been command over resources by local priority area teams (P.A.T.s),
and a strong centre has been favoured in the
belief that this provides greater flexibility
at the periphery than that of traditional
area management systems.   This emphasis on
alleviating stress rather than devolving broad
local government functions is crucial in an
assessment of the scheme and the response
to it by local organisations and community
workers.

Essentially the programme consists of the
specification of priority areas on the basis of
social indicators.   This is coupled with the establishment (in relevant wards) of priority area teams,
which are composed of ward councillors and serviced
by full time personnel.   This involvement of local
councillors is associated with the development
of public participation, both through the involvement of voluntary bodies and the holding of public
team meetings to discuss and, to some extent,
determine the use of resources allocated to the
teams.   Miller's invaluable article focuses on
the extent to which the whole intellectual basis
of the exercise involves a wilful (in the Newcastle
context the emphasis on volition is quite proper)
rejection of evidence on the structural origins
of what is described as 'stress'.   I will return
in a moment to the way in which utterly discredited
and fallacious accounts of the origins of deprivation in the 'social pathology' of collectivities
and individuals are being re-asserted as the basis
of professional practice in community work.   However, Miller also very clearly recognises the contradictory and incorporating nature of the exercise
as a political process.   As he (1981, p. 98) says:

> ...most political activists in the community
> saw it as an offer to participate in, and
> thereby construct, a programme that could
> never realistically make a substantial impact
> on the problems of the inner city.

Yet the reality is that many political activists have participated because it is almost impossible not to and one of the most important modes
of participation has been through actual employment
on schemes.   Miller is very clear about this.   As
he says:

...there is often a gap between such theoretical decisions (i.e. accurate analysis of
the nature of the game) and what actually
happens in practice. In Newcastle the city
clearly took the initiative in establishing
the programme and have maintained that position
ever since. Thus, although some groups have
attempted to offer alternative definitions
and proposals, their influence has been insufficient to affect the 'hierarchy of credi-
bility' used by large sections of the population. The authority has retained its ability
to control the terms, the forms and the content
of local political debate and action, thereby
removing important issues about unequal distribution of scarce resources from local politics
and redefining public issues, private troubles
and political issues as technical concerns.
(Miller 1981, p. 108.)

Piven and Cloward (1979) produce an account
of the incorporation of issue based 'poor people's
movements' into existing structures of power mediated through bureaucracies. Although this was
very far indeed from their intention, they might
have been writing a tactical manual for the distortion of class politics in a context of de-industrialisation. To explain why this is so, let me
return to a discussion of autonomist politics in
relation to reproduction.

Is the 'War of Position' Turning into Capital's 'War of Movement'?(3)

The core of the 'autonomist' account of capitalism
is the identification of the working class as the
moving force in change. There is an explicit rejection of 'capital logic'. As Cleaver (1979,
p. 53) puts it:

If autonomous workers' power forces reorganisation and changes in capital that develop
it, then capital cannot be understood as an
outside force independent of the working class.
It must be understood as the class relation
itself...In other words, capital seeks to
incorporate the working class within itself
as simply labour-power, whereas the working
class affirms itself as an independent class
-for itself and through struggles which rupture
capital's self-reproduction.

The background to this approach lies essentially in a discussion of the labour process at the point of production. Thus, Braverman's (1974) account of the development of management is, whatever criticisms may be made of it, an account of class conflict in which management is used to 'incorporate the working class...as simply labour-power'. But work is not everything. What about 'reproduction', an activity or series of activities conducted within civil society and in relation to the state?

Before Andre Gorz got a bad dose of ecology and reverted to his existentialist root-stock, he wrote:

> ...partial victories won in this way if they improve living conditions will not therefore reinforce capitalism. On the contrary, the public ex propriation of real estate, the socialisation of housing construction, free medicine, the nationalisation of the pharmaceutical industry, public cleansing and transportation services, an increase in collective facilities, regional development planning (elaborated and executed under the control of local assemblies and financed by local funds), and the social control of all these sectors which are necessarily outside the criteria of profit, these things weaken and counteract the capitalist system from within. Their mere functioning as social services require a constant struggle against the capitalist system itself, since they cannot be kept working without a form of social control over the whole process of capitalist accumulation and the latter's subordination to a democratically determined scale of priorities reflecting the scale of needs (Gorz 1964, p. 97).

The year 1975 seems to me to be a key date here. Until then the working class were making gains, and some of their most important gains were politically achieved, in reproduction, through a creative use of the local state, even if the creative imagination behind these gains was by 1975 some 80 years old. Since 1975, when the I.M.F. imposed the system's logic (see Harris, 1980 for a good discussion of the idea of 'system logic'), there has been a consistent attempt to change the character of local administration away

from the working class. This did not begin in
1974. Dunleavy's (1981) discussion of the attempt
at a technical fix through mass housing shows that
pressures in this direction were strong in the
1960s. Indeed in the 'war of position' the middle
1930s (see Byrne 1979) displayed something of the
same characteristics.

However the corporatist change is now so pervasive that one might think of 'capital' as having
broken through into a war of movement with poverty
professionals acting as line of communications
troops keeping the rear in order. How do they
do this? It seems to me that they do this by mis-
directing themselves and others. The system gives
results so work within it is the message. The old
political message was 'we'll take from the system
and if it can't deliver we'll change the system',
and there was some sort of idea about what it might
be changed to. There were utopias of development.
Offe has commented on the different characters
of the utopias of today, utopias often endorsed
in terms of lifestyle politics, especially
by poverty professionals. Social democracy was
for progress. Contemporary utopias are different:

> What dominates the thought and action of these
> (social) movements is not a progressive
> 'utopia' of what desirable social achievements
> must be achieved, but a conservative utopia
> of what non-negotiable essentials must not
> be threatened and sacrificed in the name of
> 'progress'. (Offe 1984, p. 190).

It is all very defensive.
Indeed to say that is defensive may be an
exaggeration. Although there is forceful resistance to the trend (see Craig, Derricourt and Long
1982) there is a very definite re-emergence of
bourgeois structural functionalism in the form
of the 'unitary' approach, as the basis, of a 'professional community work' practice. This is particularly well represented in work produced by
the National Institute for Social Work Training
and associates, most especially in David A. Thomas,
The Making of Community Work, (1983). I have not
the space here to lay out a critique, or more
properly repeat the critique, of what these moves
towards 'professionalism' on the basis of systems
theory consist. Suffice it to say that they are,
weak and discredited as they may be, an attempt
at providing an intellectual basis for

collaboration.   To be confined to the defensive
in this context is regrettable.   To collaborate
is contemptible.

Conclusion

I want in concluding to return to the positions
asserted in the introduction to this piece. Let
me do this by saying that the positions there
asserted are contrary, and to be contrary there
has to be something to be against.   In this instance
I would argue that what is contained in that introduction is opposed to a common position which is
developing in contemporary analysis - a position
that can be identified with Euro-communism in politics and with left-weberianism in sociology and
intellectual disciplines associated with sociology.
While rewriting this piece for publication I read
over work from two fields: (1) socio-political
analyses of the spatial, exemplified by a number
of recent contributions from, _inter alia_, John
Urry; and (2) reviews of the problems surrounding
community work as a series of social practices.
I was greatly struck by the congruence between
these despite an absence of inter-reference. Thus
Urry's discussion 'De-industrialisation, classes
and politics' (1983) is remarkably like Blagg and
Derricourt's account of 'Why we need to reconstruct
a theory of the state for community work' (1982).
These bodies of work are interesting
and important.   If I disagree with the conclusions
reached, and I do disagree, it is not a matter
of the polemical rejection which is appropriate
to the denunciation of the reassertion of utterly
discredited perspectives as the basis for community
work practice, but rather a matter of intellectual
and political argument.   In other words, I found
them useful but think they are wrong.   Basically
these writers, like Offe, see society as changing
in ways which remove the capital-labour relation
from its central position as the basis of division
and social action.   They emphasise gender, race
and ethnicity and conflict over issues of 'consumption' as the new frames around which social conflict will be conducted.   In other words they are
attempting, particularly through a usage of the
notion of 'civil society' to construct an analysis
of the relationship between society and politics
which makes sense in the 1980s.
Now it ought to be perfectly clear from the
foregoing that I agree with the essence of the

description attempted. Indeed a writer in the
traditions I am contrary to could well attempt
to articulate my description of Gateshead-Newcastle
as empirical support for their account. And yet
I do not agree. On one thing I do agree with Urry
(1983) and that is that space is crucial to our
understanding here - both in terms of the use of
space by capital in disorganising (underdeveloping
in the autonomist sense) the working-class and
in the importance of spatial propinquity as the
basis of the re-organisation of that class. However, I do not agree that what is happening is
separable from the fundamental, basic importance
of the wage labour - capital relation, even if
for many people this is no longer, directly,
experienced.

Urry is criticising Saunders' and Cawson's
emphasis of the 'politics of consumption' makes
the point that:

> It is surely not the case that local struggles
> necessarily revolve around the politics of
> consumption. The de-industrialisation of
> an economy effects a substantial restructuring
> of the politics of production. This is partly
> because production is put back on the pol-
> itical agenda (if ever it went off) but in
> a manner in which struggles revolve around
> the re-capitalisation of localities. (Urry
> 1983, p. 46).

This is certainly true, albeit such 'struggles'
are often conducted in a profoundly corporatist
fashion at a regional level which is decidedly
remote from the actual experience of the people
whose lives are affected. What are the experiences
such people do have? The account of the workings
of the Gateshead-Newcastle Inner City Partnership
and associated activities is not an account of
the direct experience of the 'masses'. Probably
many, if not most, of the masses in the area are
unaware of the existence of any such animal. It
is an account of experiences which are central
to the lives of those who attempt to organise on
the basis of issues of reproduction and of those
employed by the state in relation to such organisation.

I conclude that the end effect is to disorganise the working class in the interests of
facilitating the restructuring of capital. More
specifically, I conclude that this disorganisation

is more a function of structure than of process. Many of the forms of intervention associated with community work practices (and I should note that many of those engaged in this are not formally described as community workers) have great radical working class potential. However they are conducted in relation to a structure of decision-making which is a 'stranger to democratic theory'.

Let me conclude by summarising. The argument of this paper has been that in a reaction against the autonomous radical reformist project of the working class, capital in the UK has developed a series of administrative mechanism's best defined as 'neo-corporatist' which are directed at managing the crisis which emerges from the disjunction between radical reformist objectives; in reproduction and the system logic of contemporary accumulation. These work by 'depoliticising' or more precisely 'de-democratising' the processes of policy formation and implementation. A specific example is provided by the workings of the 'Inner City Partnership' in Gateshead-Newcastle which deals in large part with the management of the inner-city 'marginalised'. The relation of administration to class is mediated through a set of 'poverty professionals' who in day to day interaction impose the logic of the system on potentially disruptive groups of individuals. The question is how to stimulate and develop a progressive programme and movement which might be the basis of the counter-attack in what is a very depressing situation.

Since the disorganisation is reducible to the nature of the wage labour - capital relation, then these struggles will be implicated in that relation. They are neither less nor more important than class struggles at the point of production in principle. Tactically they may currently be more important but they cannot be separate. Fundamentally these are class struggles too.

NOTES

1. Unlike Offe I still endorse the notion that 'base determines superstructure' provided that we follow Williams's (1982) redefinition of every term in the proposition.
2. I am quite old-fashioned enough to believe that the vanguard in changing what is to what might be, at least in part, is drawn from that world. I am not bidding 'Farewell to the Working Class' (Gorz 1982).

3. These terms are Gramsci's. In general he saw the conflict between classes in western capitalism in terms of a war of position, both in civil society and in the spheres of production.

REFERENCES

Anderson, J. (1983) 'Geography as ideology and the politics of crisis: the enterprise zones experiment' in J. Anderson, S. Duncan, R. Hudson (eds.), Redundant Spaces in Cities and Regions, Academic Press, London, pp. 313-350.

Bennington, J. (1975) Local Government Becomes Big Business, CDP, London.

Blagg, H. and Derricourt, N. (1982) 'Why we need to reconstruct a theory of the state for community work' in G. Craig, N. Derricourt, N. and M. Loney, (eds.), Community Work and the State, Routledge and Kegan Paul, London, pp. 11-23.

Braverman, H. (1974) Labor & Monopoly Capital Monthly Review Press, New York.

Byrne, D.S. (1979) 'The decline in the standard of interwar council housing in North Shields' in J. Melling (ed.), Housing, Social Policy and the State, Croom Helm, London, pp. 168-193.

Byrne, D.S. (1982) 'Class and the local state', International Journal of Urban and Regional Research, 6, 61-82.

Byrne, D.S. and Parson, D. (1983) 'The state and the reserve army: the management of class relations in space', in J. Anderson, S. Duncan and R. Hudson (eds.), Redundant Spaces in Cities and Regions, Academic Press, London, pp. 127-154.

Cleaver, H. (1977) 'Malaria, the politics of public health and the international crisis', Review of Radical Political Economics, 9, pp. 81-103.

Cleaver, H. (1979) Reading Capital Politically, Harvester, Brighton.

Cockburn, C. (1977) The Local State, Pluto Press, London.

Craig, G., Derricourt N, and Loney, M. (eds.) (1982) Community Work and the State, Routledge and Kegan Paul, London.

Dunleavy, P. (1981) Mass Housing in Britain, Clarendon Press, Oxford.

Friedman, A. (1977) Industry and Labour, Macmillan,

London.

Gateshead Director of Planning (1983) Quarterly Unemployment Review March 1983, Gateshead Metropolitan Borough Council, Gateshead.

Gorz, A. (1964) Strategy for Labour, Beacon Press, Boston.

Gorz, A. (1982) Farewell to the Working Class, Pluto Press, London.

Green, D.G. (1981) Power & Poverty in an English City, Allen and Unwin, London.

Harris, L. (1980) 'The state and the economy: some theoretical problems' in R. Miliband and J. Saville (eds.), The Socialist Register 1980, Merlin Press, London, pp. 243-262.

Miller, C. (1981) 'Area management: Newcastle's priority areas programme' in C. Smith and D. Jones (eds.), Deprivation, Participation and Community Action, Routledge and Kegan Paul, London, pp. 97-111.

Newcastle-Gateshead Inner City Partnership (1983) The Inner City - within Newcastle and Gateshead, Tyne and Wear County Council, Newcastle.

Newcastle-Gateshead Inner City Partnership (1984) Action Programme 1984-1987, Tyne and Wear County Council, Newcastle.

North Tyneside CDP (1977) Final Report (four volumes) North Shields.

O'Connor, J. (1981) 'The meaning of crisis', International Journal of Urban and Regional Research, 5, pp. 301-328.

Offe, C. (1984) Contradictions of the Welfare State, Hutchinson, London.

Panitch, L. (1980) 'Recent theorisation of corporatism', British Journal of Sociology, 30, pp. 36-51.

Piven, F.F. and Cloward, R. (1979) Poor People's Movements: How They Succeed and Why They Fail, Vintage Books, New York.

Seabrook, J. (1984) The Idea of Neighbourhood, Pluto Press, London.

Shutt, J. (1984) 'Tory enterprise zones and the labour movement', in Capital and Class, 23, (Summer), pp. 19-44.

Thomas, D.N. (1983) The Making of Community Work, Allen and Unwin, London.

Urry, J. (1983) 'De-industrialisation, classes & politics' in R. King (eds.), Capital and Politics, Routledge and Kegan Paul, London, pp. 28-48.

Williams, R. (1982) 'Base and superstructure in marxist cultural theory' in R. Williams,

Problems in Materialism and Culture, Verso, London, pp. 31-49.

CHAPTER SEVEN

CLASS RELATIONS AND LOCAL ECONOMIC PLANNING

JAMIE GOUGH

1. Introduction

This paper examines some aspects of the role of
local authority economic intervention (Local Government Studies 1981, Young and Mason 1983). In the
reproduction of class structure and class relations.
Some brief comments should be made on this choice
of object. Firstly, there is a presumption that
the activities of the state can play an important
role in shaping class relations: the latter are
not simply given by structures established at the
level of production, though these are the theoretical starting point. One can make an initial
distinction between two ways in which state economic
interventions have an impact on class structure.
Firstly, and most obviously, these interventions
alter production itself; secondly, they both given
expression to, and shape, the political organisation
of the classes. However, it would be misleading
to regard these two aspects as substantially separate levels of intervention, a 'material/economic
level' and a 'political/ideological level', or
even as substantially separate effects: the two
aspects are in reality closely intertwined. Changes
in production instigated by the state nearly always
involve qualitative changes (in labour processes,
in ownership structures, etc.), and these alter
the basis on which class organisation is built.
Moreover, the very process of carrying out these
changes involves the activity of more or
less organised sections of classes, and this
activity reproduces and changes those class
organisations. The place of ideology is
equally complex: it operates not only through
political organisation of the classes, but is also
simultaneously an aspect (a 'moment') of production

practices (labour processes, divisions of labour, etc.).

These comments are relevant in considering the form and effects of recent local authority economic interventions. It is tempting to regard these interventions as essentially ideological. Their size is extremely small when compared to that of the economies into which they are intervening: for example, the programme of the Greater London Council, one of the most ambitious in the country, has a budget of around £30 million per annum, compared with output of the London economy of the order of £40 billion. The simple material impact of the intervention is therefore necessarily small (like 'trying to water London's gardens by spitting from a helicopter', as one industrialist described it). Moreover, many local authorities seem to make up with hype what they lack in material resources: a high proportion of funds for local intervention is spent on advertising, promotion and publicity of various kinds.

The ideological aspect of local economic planning (LEP) is undoubtedly important, and will be a major theme of this paper. But a notion of this intervention as a ideological practice is an inadequate, and ultimately misleading, one, even for analysing the ideological aspects of that intervention. In the first place, there is typically a major disparity between the ideology of these interventions, in the sense of the way in which they present themselves, and their reality, the actual practices they involve. More fundamentally, as suggested above, their ideological content is best seen as an aspect of social practices, both in production and in class organisation, which also have a material aspect. The most ideologically potent forms of local intervention are not the overtly ideological pronouncements of the participants in those interventions, but rather are those which are embodied in determinate social practices. Local authority interventions are small scale; but they may play a part in setting up qualitatively new forms of political-economic relations which may have a material impact beyond the parts of the economy directly affected. This is recognised, in a certain sense, by both the left (in the notion of 'exemplary projects') and by the right (in 'experiments' such as the Enterprise Zones).

This way of approaching the problem also allows us to analyse the contradictions in the ideologies of LEP. This is so in a double sense. The diversity of practices involved in a local economy on the one hand and in LEP on the other will be associated with a diversity of ideologies which are typically mutually incompatible and divergent; and these ideologies in turn typically differ significantly from the official presentation of LEP(1). The analysis of this paper therefore attempts to include the self-presentations of policy, the reality of the policies (the social practices they embody), and the relation between these two. In this way one can hope to grasp of what the policies are really 'exemplary'.

This approach is also helpful in considering an important issue within LEP, which will be a theme of this paper: the relation of local economic strategies to national policies. This relation has been stressed by both the left and the right in carrying out local interventions. Left Labour-controlled councils have presented their inter-ventions not only as dependent on sympathetic cen-tral government action in order to expand, but also as exemplary of policies that might be carried out at a central level (that is, a qualitative as well as a quantitative relation). The present government has presented its own local inter-ventions, particularly the Enterprise Zones, as experiments for a national policy. At the level of overt ideology, there are indeed rather obvious congruences between many of the policies being pursued within LEP and policies put forward by both right and left for the national economy. The commitment of both the present government and many local authorities to small firms and the arguments used to justify this commitment, are one example. But our approach here indicates the need to go beyond this observation, to examine the real practices of LEP and the ideologies that these <u>practices</u> embody. This opens the possibility that even the ideological content of LEP may differ from that of apparently similar policies (actually or potentially) carried out at a national level.

Of the many axes of class formation that might be analysed in relation to LEP, this paper focusses on just one, which however appears to me to be fundamental: the extent to which LEP promotes, and is based on, independent activity and organ-isation of workers. I therefore examine to what extent LEP, in its effects, strengthens or sup-presses action by workers with aims which are dis-tinct from those of capital, and using organisation which is separate from capital (the trade unions in the first place). As we shall see, this action affects not only the relation of the working class to capital, but also crucially conditions relations between different sections of the working class.

Closely related, but conceptually distinct, is the question of how LEP is produced and implemented. State economic intervention in Britain has typically been developed either without the participation of workers' organisations, or (for example in the period of tripartism from the early 1960s to the late 1970s) with the participation of the trade union leadership. The latter arrangements were aimed not at stimulating the development and expression of a distinct policy on the part of the workforce, but on the contrary at the production of a consensus, an equivalence, between the positions of labour and capital. Not coincidentally, the participation from the trade union side in this state planning was usually limited to a narrow layer of the union leadership. Below I examine some of the barriers to LEP breaking out of this traditional pattern.

A word should be said about the method by which LEP is analysed in this paper. My starting point is an assumption of certain aims for LEP, related to the issues just discussed (for example, an aim of involving workers in production of their own economic plans); the barriers to realising this aim are then examined. This method might be read as containing the theoretical implication that state policy fundamentally originates in the ideas and aspirations of state personnel. No such implication is intended. Rather, this method is used to focus the discussion in a particular way. Firstly, the aim of the paper is not to analyse the origins of current LEP but rather to analyse its forms and effects. Secondly, the method adopted enables the discussion to be more speculative, going beyond existing initiatives. This seems appropriate given the fact that the types of LEP examined here are relatively new and under-developed(2).

Final introductory points concern the concrete object of discussion. I concentrate in this paper on LEP of the type being carried out by a number of left labour councils ('left LEP') (Boddy 1984). These initiatives present themselves as setting out to construct a new relation between the local state and labour and to promote a new role for workers within economic planning. They are therefore particularly relevant to the theoretical

concerns already mentioned. The empirical material
on which the discussion is founded is largely taken
from the experience of local economic planning
at the Greater London Council (GLC) and its investment arm, the Greater London Enterprise Board,
over the last three years; this evidence is not
however set out in the paper (GLC 1985a). The paper
is not intended as an assessment of (even a part
of) the GLC's LEP to date, but is rather a reflection on some of the experience which has
been gained.

A considerable variety of instruments is now
used by LEP. I shall focus on just one of these,
the provision of funds to firms, in the form of
grants (e.g. premises related grants under the
Inner Urban Areas Act) or loan or equity finance
(e.g. Enterprise Board funding, IUAA loans). The
reason for this focus is that within LEP this investment function raises some of the most difficult
problems from the point of view of the social relations considered here. In this, it may be con-
trasted with a number of activities being undertaken by the GLC and other local authorities which
give direct support to the organisation of labour:
the financing of trade union support centres, of
local trade union campaigns, of international
meetings of workers within multinationals, for
example. While the latter mode of economic intervention raises issues about labour movement autonomy
and democracy, it does not start with the inherent
ambiguity of the investment function: accumulation
of capital in the interests of labour. This ambiguity tends to mean that it is within the investment
function that the class relations of LEP are most
sharply brought to light.

Even within left local investment practice,
there are an enormous variety of approaches and
focuses. Because of this, the discussion below
is far from comprehensive, but discusses a range
of different policies in an exploratory and preliminary way.

The main stages of the argument may now be
summarised, and this provides a plan of the remainder of the paper. First, it is argued that the
tendency of LEP as practiced by most local authorities in Britain is to encourage and use workers'
organisation only insofar as this backs the
interests of capital operating in the locality.
The question is therefore posed as how this
'localism' can be avoided. The question of localism
is posed most sharply in relation to quantity of

of employment; section three of the paper therefore examines the issue of job creation. It is concluded that, to the extent that the aim of LEP is job creation, there is little scope for a policy that is not localist. Successive sections therefore examine the implications for class relations of policies which, at least in part, set aims other than job creation. In section four the possibilities and problems of progressive reforms in discrete areas of policy are discussed. In sections five and six a critique is given of two overall strategies for sectoral restructuring, 'strategic market planning' and the revival of the industrial district. Section seven looks at the implications for class relations of firm-by-firm intervention as an alternative to sectoral investment strategy.

2. Class relations and mainstream local economic policy

In order to bring into sharper focus the problems of left LEP, it is helpful to consider briefly the social relations of 'mainstream' LEP as practiced by the majority of local authorities, New Town Development Corporations and Development Agencies. These LEPs typically combine policies aimed at attracting inward investment (financial incentives of various kinds, information, advertising) with policies aimed at preventing outward movement of jobs from their area. They do this by attempting to meet the needs of various targetted sectors of capital (for good premises, skilled labour, bridging finance, etc.), and also by advertising aimed at convincing businessmen that their area offers other attractions, not necessarily provided by the local state agency itself (good intranational communications, a compliant and 'available' workforce, etc.). To the extent that labour in the particular area benefits from this activity, in the form of increased job opportunities, it is by virtue of the advantages to capital provided by the LEP or the market.
  Labour is thus invited to see itself in a particular way. Firstly, its interests coincide with those of capital which potentially or actually operates in its area (which we may call 'local capital'). Labour furthers its own interests either by remaining passive, or by allying with local capital to lobby the local or national state for more resources to be put at the disposal of local capital. In encouraging this type of alliance,

173

LEP continues the class politics of traditional regional planning. The alliance is often given organisational form in tripartite bodies of various kinds, providing a forum for the development of an economic policy 'in the interests of the local area' and for organising joint lobbying of central government for powers and funds to carry this out.

Secondly, workers in the locality are invited to see their interests as directly opposed to those of workers outside it; jobs are to be got by effective competition with enterprises in other locations. In fact, the point may be posed more strongly; this competition is typically more sharply competition with labour elsewhere than with capital elsewhere. Where LEP seeks to strengthen indigenous firms relative to those outside the locality, the losers are both capital and labour in other areas. But where LEP seeks to attract capital from other locations or mobile capital not yet anchored to a particular location, the losers are only workers in other locations. For mobile capital, the spatial competition promoted by mainstream LEP provides opportunities rather than problems.

Spatial competition, and its attendant class relations, may therefore be seen as the overarching strategy of mainstream LEP. But this strategy operates through a number of quite varied tactics and mechanisms: policies on premises, training, new technology, infrastructure, etc. The remarks I made in the introduction might prompt the question: are the class relations promoted by these particular mechanisms compatible with those of the overall strategy? I think that in general the answer to this question is 'yes', but I can here only give one illustration: the widespread promotion within mainstream LEP of small firms as 'the employment creators of the eighties'. Local authorities have tended to rationalise the fact that their resources stretch only to helping small firms by adopting various themes of the small firm mythology (GLC 1983). The themes relevant to us here are those concerning the relation between employer and workers within the small firm. Small firms are said to be more 'adaptable', more 'flexible' and more innovative. While these ideas deserve the label 'mythology' (for, amongst other reasons, that they create a falsely homogeneous category 'small firm'), they do give euphemistic expression to what are realities in a large proportion of small firms: poor or non-existent organisation of labour, lack of formal controls on health and safety, lack of job descriptions, informal or non-existent contracts of employment, compulsory over-time to meet order deadlines, and so on. Since mainstream LEP typically promotes small firms as such, discriminating only according to their growth prospects, firms of this type are included along with ones offering better employment conditions. The way in which the firm is 'flexible' etc. is thus not problematised or contested: if this is at the expense of labour, that is not relevant to 'economic planning'. The promotion of small firms as such therefore contains within it an identification of the interests of labour with those of local capital and a subordination of labour's organisation to the needs of local capital. This is congruent with the social re-lations of spatial competition.

But this is not all. The class relations pro-moted through this type of LEP fit rather closely those which we have already noted as characteristic of central government strategy since the 1960s. The identification of the interests of capital and labour in a competitive struggle against those elsewhere (in this case, in other countries), the presumption that collectively labour can best further its interests by passivity, and the need to subordinate the quality of work and labour organ-isation to capital's need for 'flexibility', are recurring themes in the practice of national econ-omic policy. The present government has shifted the emphasis and articulation of these themes but not fundamentally altered them(3).

We can therefore conclude that in terms of class relations there is a congruence, at least at the level of abstraction considered here, not only between the strategy and tactics of mainstream LEP but also between these and traditional forms of central government economic intervention; 'localism' in this regard can be quite compatible with national economic policy. This is not, of course, to argue that there are no contradictions in this type of LEP, merely that they do not appear at this level.

The GLC and some other left authorities have (though not always consistently) presented their local economic interventions as an alternative to 'localist' LEP and the class relations it in-volves. In terms of class relations, the stated aim is, firstly, that the interventions should promote labour as an economic subject, encouraging and enabling workers to put forward their own plans

175

for their firm, industry or local economy. In relation to the mode of operation of the local state, the aim is that the local authority should at least consult closely with labour in the development of its plans (the participative state), and where possible promote and give resources to workers' own initiatives (the state as an instrument of organised labour).

The question of this paper may therefore be re-posed: to what extent can the stated aims of a left local investment strategy with regard to class relations be achieved, and to what extent can such a policy break from the class relations of 'localist' LEP? This question is posed most sharply on the terrain of jobs creation, and we now turn to this issue.

3. 'Employment creation': the numbers game and social relations

The centrality of the problem of unemployment to both the unemployed and the employed has meant that the provision or retention of jobs has generally been professed as the first aim of all local economic interventions. The GLC has distinguished its approach from mainstream LEP in that it does not aim systematically to attract production from outside its area, particularly from other areas of high unemployment. For employment to be created in a genuinely non-competitive fashion, however, interventions are needed that can produce a net increase in the number of jobs not just in the locality but in the national/international economy as a whole. If, on the other hand, intervention is necessarily at the expense of workers elsewhere, then the effect as far as this aim of intervention is concerned is that of mainstream LEP - to construct the interests of workers as antagonistic to those in other industries or locations. We therefore need to ask whether net job creation is possible via LEP, and whether job creation initiatives can be carried out in a way which is consistent with the class relations proposed by left LEP.

Aside from simply asserting that jobs created or retained within a particular enterprise attack unemployment, a number of arguments are used to show that net job creation by LEP is possible:

(i) One type of argument locates net job creation in switches between particular types of sector(4). It is argued that investment in a new product or new product area does not compete with production, and therefore jobs, elsewhere. Product innovation thus becomes the key to net job creation. But LEP cannot influence aggregate demand (whether as a whole or in its main divisions of consumer goods, investment goods, public sector purchasing, etc.); there is therefore inevitably diversion of expenditure from other products. A second form of this argument is that switching investment or expenditure from capital intensive to labour intensive production can create jobs. But this is to neglect employment in the production of capital goods used in these sectors of production.

(ii) A more sophisticated argument locates the job creation role of LEP centrally in the accumulation of capital. It is argued that the role of local enterprise boards should be to put together profitable investments that would not otherwise be assembled. This need not merely divert funds from other investments, since it can mobilise funds from productive (employment creating) investment in a way that would not otherwise have occurred. This happens either through mobilising idle capital or capital that would otherwise be tied up in unproductive activities (e.g. speculation); or through the banks creating more money in response to the availability of new profitable investment opportunities. In other words, LEP can inject new funds into the local economy, can reflate it, in the absence of powers over monetary control or deficit financing, via production itself (Murray 1983a).

This raises the question of how LEP is to help create such profitable investments. Some of the agencies set up for local economic interventions (Enterprise Boards, business advice centres, cooperative development agencies etc.) do identify investment projects and market niches which might not otherwise have been perceived. But this merely adds a few more entrepreneurs to the pool. For intervention by the local state to be justified, it must be shown that this intervention can systematically play a role that would otherwise not be fulfilled.

The classic means by which the market raises profitability and creates new investment possibilities within sectors and the economy as a whole is via, on the one hand, devalorisation of capital (bankruptcies, scrapping of capacity, etc.), and on the other hand lowering of the share of wages in value added. The state can promote these processes; this in fact is the essence of monetarist policy. But this is clearly creating profitable investments at the expense of jobs or wages or both.

A second systematic reply to the question of how new profitable projects are to be generated is that this is to be done by improving the quality of management of enterprises. The argument is based on the premise that British management, particularly that in manufacturing, is poorly trained for its role and often incompetent, and that this has been a central cause of low profitability in the UK economy(5). Local state intervention can change this, by replacement of management personnel, carrying out of some management functions within local enterprise boards, management training schemes and so on.

The premise of this argument seems to me false, or more precisely superficial. Historical analysis of the weakness of British manufacturing indicates that the poor quality of British management is essentially an effect rather than a cause of the UK's relative industrial decline (Glyn and Harrison 1980; Gamble 1981). In any case, the effects of a strategy of 'improving management quality' on class relations run against the stated aims of left LEP. Firstly, the argument is a parochial one: the policy has the aim of bringing UK practice up to the standard of other advanced capitalist countries. Within LEP, the aim may be to implement this strategy within a particular locality; or it may be given even stronger parochial content where it is argued that it is particular areas of the UK such as the cities or 'the inner city' which are lacking in management and entrepreneurial talent, and that the role of LEP is to make up for this lack. Whatever the variant, we are returned to the politics of 'localism'. Secondly, the strategy gives no role to the collective organisation of labour. At most, it may seek to use the ideas of individual workers for changes in products or processes ('tapping the gold in workers' minds', Japanese style). But in general, even individual workers do not have the training, resources or the time to beat management of competitor firms at their own game, in any systematic way(6). Improvements to management will be made essentially at the level of management itself, and with the state operating autonomously from the organised workforce.

(iii) An alternative argument is that the local state can play a role in reflating the economy by making funds available to industry on better terms than would otherwise be the case. This argument relies on a premise which is directly contrary to that in the argument just examined. Rather than locating stagnation of the economy in an absence of profitable investment opportunities, it sees stagnation as a result of inadequate funding for existing (potential) profitable investments (Minns 1982, 1983). This inadequacy arises from the peculiar nature of the British financial system, which has meant that it has set much more stringent conditions on risk, and much shorter repayment periods for loans, than financial institutions in other leading capitalist countries. I would assert that this again mistakes cause for effect: the attitude of the City to British industry is fundamentally the result, not the cause, of British industry's chronic low profitability, though the latter has undoubtedly been reinforced by the historically acquired structure of British capital (Murray 1983b).

But again, it is the implications for class relations that are important here. The problem, and thus the strategy, again revolves around supposed peculiarities of the UK and its competition with other countries. Within LEP, this can be reinforced by arguments which locate local weakness of industry in local difficulties of access to finance(7). This strategy sees the local state as acting on behalf of both labour and capital within the 'productive' sector of the economy, to mobilise funds (from rate income, from pension funds, etc.) which would otherwise be invested unproductively or be exported. The trajectory of this policy at a national level, even in its most radical version, is an alliance of 'productive' capital and labour against the City(8). An independent politics of labour is once again absent.

We may conclude that there are severe barriers to LEP being a net job creator(9). Local authorities lack the power of central government to operate reflation via monetary policy (whatever the limitations of such policy, including its lack of direct purchase on production); and their powers in the sphere of production fall far short of the extra-market measures (appropriation of assets, etc.) that could directly mobilise under-used resources. Thus, while a local intervention may set out to provide a direct example of how a national government could combat unemployment, it is deflected from this aim by the limited power of local government(10).

The real effects of a job creation strategy are different. If positive employment effects are at the expense of jobs elsewhere, such strategies are inevitably localist (calculations and comparisons of 'cost per job created' are therefore otiose, or should be rephrased as 'cost of job transferred'). Moreover, we have seen how the bipartisan politics of localism are both reinforced and given a particular inflection (new-technological, pro-services, entrepreneurial, managerialist, anti-City) by the particular policy used for job creation. In doing so, they undermine the development of an independent strategy for labour which, in breaking from the problematic of competition, could propose a viable strategy for job creation (that is, for a planned economy, see footnote 9).

This does not mean that all versions of the employment zero sum game are equivalent. For example, the preservation of jobs in situ in competition with enterprises elsewhere has different political implications from encouraging moves of production from one area to another. The point is, rather, that it is these political implications which are crucial, rather than some claim that jobs have been 'created'.

This conclusion implies that the professed aims of left LEP require the abandonment of ambitions and ideology of net job creation, and a concentration directly on qualitative questions. The competitiveness of enterprises with which the local authority is involved does not, of course, become irrelevant(11), merely that this is seen as a means to achieving other ends, and therefore as subordinate to the latter. Rather than 'social policies' being the icing on the cake of a strategy for increased competitiveness(12), changing social relations becomes the aim of intervention.

4.   Class Issues in a Sectoral Approach

Some ways in which such intervention might be carried out are discussed in the next sections. The focus is on interventions organised on a sector basis, rather than one-off initiatives. This sector form of intervention ('sector planning') has been adopted by a number of left Labour local authorities (Greater London Council 1985a). One motivation for this is that a sectoral approach may facilitate the development of relations between the local authority and the unions, especially in sectors with an industry union. This implies a definition of 'sectors' which correspond to the organisation of labour, rather than to product markets for example(13). But what can be the aims of such a sectoral policy?

One issue that such a policy can address is geographical unevenness in the strength of labour organisation, and the way in which this is used by capital. An example would be an industry where well-organised labour in the locality was competing against poorly organised labour elsewhere, typically as a result of a more or less conscious strategy on the part of capital to transfer production from the first to the second (from a conurbation to a small town or rural area, for example). An intervention which strengthened the competitiveness of the local sector, while 'localist' in form, would have the effect of strengthening or defending the organisation of labour against capital. Another, less common, example would be in a growth sector, where growth of the industry in the locality would be on the basis of decisively stronger labour organisation than if the growth took place elsewhere. LEP operated on this basis would constitute the defence of the organisation of labour by the state, albeit in a rather indirect fashion. The problem is then posed at another level, as the classic trade union question: in what sense is it in the interests of labour as a whole to defend strongly organised labour against replacement of weakly organised labour? - a question that will not be taken up here.

The defence of strong labour organisation by LEP need not necessarily have a spatial aspect. In setting the priorities for local intervention, sectors or subsectors where conditions of employment are generally poor and labour organisation weak can be targetted. These are often also sectors in which women and black people are ghettoised (see section on printing in GLC 1985a). The aim of the intervention would be, via partial or total state ownership of some enterprises, to allow and encourage stronger labour organisation and the negotiation of better conditions of employment; the fact that these were shown to be possible in the sector in question may then encourage workers in other enterprises to press for similar conditions. The interventions could thus act as benchmarks for collective bargaining - provided, of course, that the enterprises were able to survive with the improved conditions of labour. One of the dangers of this strategy, however, is of it remaining intervention on behalf of labour rather than with labour, since by hypothesis workers' organisation in these sectors is weak. Unless intervention is carried out with, or swiftly elicits, demands on the local authority from workers in the sector, it is liable to give rise to 'stronger organisation of labour' which is formal rather than real.

Another aim that can be set for local sectoral planning is the promotion of production technologies that enable better jobs than would be produced via the market. Design of production technologies in capitalist societies involves not simply consideration of 'efficiency' in the abstract, but consideration of how the technology will enable management to better control its workforce or, by deskilling the work process, enable the use of less differentiated and cheaper labour (Braverman 1974). Local state intervention can aim to oppose this tendency, by promoting investment in technologies that maintain or build on existing skills, that allow the largest degree of discretion to the operator, and that do not pace or spy on the operator. This may involve a choice between different systems available on the market, or the promotion of the development and production of new systems.

This type of intervention can help to oppose the widely held idea of technology as an asocial product of 'pure science', entering into society like a part of nature (- a central element of the dominant ideology of microprocessor based technology in particular). It can promote the recognition of production technology as (in part) a product and an element of social power, and thus susceptible to contestation. However, paradoxically, this strategy runs the danger of leading to a different form of technical determinism - the idea of the 'technical fix'. To the extent that the strategy is carried through by the power of the local state via its control over investment funds, the strategy can come to hinge on the specification of the technology, and the ingenuity of the technologist in designing it, in detachment from the organisation of labour which it is supposed to strengthen. A pressure for this to happen arises from the fact that it is in many ways easier for the local state to relate to a small number of industry experts and technologists than it is to be part of

182

a dialogue between these and groups of workers within enterprises or sectors. The state can simply become the promoter of particular pieces of equipment, or the provider of resources to technologists to develop new products. The social relations involved may then differ little from those of a conventional science part or mainstream promotion of 'high technology', even though the technologies are said to be for the benefit of labour. Moreover, a de facto 'technical fix' approach is liable to run into the practical problem that, while technologies may constrain the organisation of work, they never determine it: a given plant can always be used in a variety of ways, with different degrees of discretion for the operator, different intensity of work, using different skills, with different division of labour. The options used will depend on the strength and form of workers' organisation. From both a practical and an ideological point of view, then, the effectiveness of a sector technology policy depends on setting up a close relation between the development of the technological options and the appropriate organisation of labour.

5.   The Local Authority as Sector Coordinator

So far we have considered only aspects of a sector intervention, in abstraction from the overall dynamic of the local industry. Some sectoral LEP has tried to develop a more inclusive approach, within which progressive changes in social relations are closely related to overall restructuring of the sector(14). A premise of these approaches is that it is possible, and indeed necessary, not simply to make reforms in particular aspects of a local sector (e.g. production technology), but to pose an alternative overall strategy for it(15). In this and the next section I examine the implications of two policies of this type for class relations.

One such approach may be termed 'strategic market planning' (Best 1984). In Britain, competition within industries is typically regulated by the market in a direct way. In contrast, in many other advanced capitalist countries market competition is mediated and modified by various forms of non-market coordination between firms. This can take place for example through banks as investors in the sector (Germany, the US at certain periods); through state direction (Japan, France); through trade associations, or less systematically, through a web of informal bilateral contacts ('the Third Italy'). These forms of coordination have been mechanisms for long term planning, partial cartelisation of markets, and partial firm collaboration in such areas as R & D and training. It is often argued that British industry's competitiveness has suffered from absense of these forms of 'planning'. A quite different trajectory for a local industry may therefore be stimulated by local government adopting a broad range of initiatives within the sector, including the resourcing of new institutions to be used collectively by local capital (funding institutions, marketing and research agencies, training centres, etc.), and by encouraging collaboration between firms.

Whatever the competitive merits of this strategy, there are problems with the social relations which it involves. Let us consider the inter-firm coordination strand of the strategy (other aspects discussed in section three). Inter-firm coordination of this type involves the discussion of business plans and the partial, controlled release of business secrets by each firm management. But the discussion is not fully public. For example, coordination often proceeds principally through bilateral discussions between the state or major bank on the one hand and the industry firms on the other. An LEP of this type is carried out by confidential discussions between, at most, the local state and a number of local firm managements, banks, etc., with the workforce excluded (except perhaps for union officials on a confidential basis).

There are reasons for this exclusion. The first is that, while strategic sectoral planning atenuates interfirm competition, it does not remove it. To the extent that the sectoral strategy developed marks out a common strategy for firms (e.g. a new subsector to enter, new production technology to use), each firm must still be concerned to differentiate itself from the others. To the extent that the strategy involves division of markets and other non-aggression pacts, these will be adhered to only in so far as it suits the survival of the particular firm (for a general discussion see Mandel, 1978). The chronic low profitability of UK industry, and the comparatively large sacrifices that would therefore often be required in strategic market planning, may be one reason why this type of planning has been weak in the

UK (suggested by Stout 1979 in his account of NEDO).
This conditional quality of the interfirm
cooperation imposes the need for secrecy.
The second, related, aspect of the problem is that
strategic market plans have an uneven impact on
the firm workforces. Under present conditions
this impact is often negative. Workers are likely
to be less ready to reconcile themselves to job
loss which appears to result from a sectoral plan
rather than from 'natural' market mechanisms. In
other words, the sectoral plan may politicise the
process of restructuring. Management's usual un-
willingness to provide information to its workforce
is thus reinforced, providing further reason for
managements to require secrecy within this type
of sectoral planning.
    To overcome these problems, deriving from
unevenness and competition between enterprises,
would seem to require a system of public ownership
and control of the industry, when the interests
of managements of individual enterprises do not
so sharply conflict with the pursuit of overall
economic efficiency. For the local state to act
as initiator and organiser of sectoral coordination
while ownership remains predominantly private tends
to both act against worker involvement in sector
planning, and to promote a closed system of planning
between state and capital. Whatever the intention
of the state, this must tend to reduce the aims
of local sectoral planning to those of local capital.

6.   The Industrial District Revisited

The ideas of strategic market planning have been
developed in a particular way within local sector
strategy around policies for the revival of the
urban industrial district. Weak varieties of such
strategies can be found within mainstream   LEP
(science parks, the Motherwell food park, some
small unit developments for particular industries),
but have been developed in a stronger form on the
left (see 'cultural industries', 'scientific instruments' 'furniture', 'clothing' in GLC 1985a;
Zeitlin 1985), and it is the latter that concern
us here. The revival of the industrial district
has been proposed as a strategy for both capital
and consumer goods industries; for brevity, we
shall consider here only the latter.
    The strategy is based on an analysis that
there are far-reaching structural changes,

encompassing both structures of production and consumption, currently taking place internationally in a large range of sub-sectors though not all (Piore and Sable 1984; Sable and Zeitlin 1985). While these changes take widely varying forms, certain general tendencies may be discerned within the consumer goods sectors:-

(i) demand, a decline in the demand for standarised goods, the typical product of the 'consumer society' of the 1950s and 1960s, and an increase in the demand for differentiated, high design, high fashion and, to some extent, higher quality goods.

(ii) strategies of the retailers/distributors (including the media majors): away from price competition in the mass market, towards selling of smaller volumes of differentiated products on the basis of quality; targetting of sub-markets by the creation of distinct images and styles, though increasingly coordinating this image across goods not only within a particular sector but across sectors.

(iii)design: design becomes the cutting edge of competitive strategy; retailers expand their design departments, but also require greater design input from their suppliers.

(iv) production: away from mass standardised production ('Fordism'), towards smaller runs, with production closely and flexibly linked to design; away from dedicated machinery and semi-skilled labour towards the use of multi-purpose advanced microprocessor-based production equipment and skilled labour; specialisation in sub-markets, including in intermediate goods, in order to obtain greatest advantage from design expertise and market contacts; hence a tendency to fragmentation of production and of ownership; requirement for skilled and experienced labour puts upward pressure on wages.

It is argued that local sectoral planning must not only take account of these tendencies since they are inexorable, but should actively promote them since they can lay the basis for industries with good wages and conditions in urban areas of the advanced capitalist countries. The tendency towards quality competition can avoid the pressures towards wage cutting and attacks on labour deriving both from the recession and form Third World competition in manufactures. The new importance of links of production to design and marketing, of access to a pool of skilled, specialist labour, and of intimate knowledge of competitors' strategies, give new advantages of agglomeration - a revival of the external economies of the traditional industrial district. They thus provide a basis for countering the de-urbanisation of manufacturing characteristic of most advanced capitalist countries.

Policy for local intervention to realise this potential characteristically has a number of elements (see GLC 1985a; Zeitlin 1985):

(a) The strategy is centred, not on the productive agents as such (retail firms, designers, producer firms, direct workers, etc.) but on their inter-relations, the market and extra-market connections between them: it is, so to speak, the external economies themselves that are to be created. Local authority resources are therefore to be concentrated in (i) intermediate institutions, such as agencies for linking designers to producers and marketing agencies, showrooms or distribution companies for groups of producers; (ii) training agencies, for managers, indirect and direct workers; and (iii) service institutions such as those discussed in the last section (for funding, for R & D, etc.). The strategy is thus a particular development of the notion of 'coordination' within 'strategic market planning'.

(b) Intervention into the producers themselves is considered problematic, since the state lacks the detailed knowledge which are the basis of competition, and may interfere with the required 'flexibility' of the enterprise. This is in contrast to the position of state intervention into Fordist producers, where the state possesses resources which may be crucial such as large scale finance and guaranteed markets. Intervention into producers would be undertaken primarily to provide an example of flexible specialisation where such is lacking in the local sector (see 'Furniture' in GLC 1985a), but would be considered undesirable if undertaken more widely.

(c) The retailers/distributors have an apparently dominant position in many consumer goods sectors, by virtue of their greater concentration of ownership compared with the producers and, increasingly, their possession of images/styles giving them control over particular sub-markets. But, even if the powers were available, it would be unnecessary and self-defeating for the state to attempt to influence the policies of these companies, particularly in their contracting arrangements. In particular, to attempt to coerce the retailers into forming long-term contractual relationships with the producers would damage the formers' competitiveness (and limit consumer choice); and this is unnecessary since the producers cannot be exploited by the retailers because they have something the retailers require - design linked to flexible production.

This kind of strategy thus has great practical attractions for local authorities: it avoids the need for control over the commanding heights, the retailers (or rather, theorises that they are not so); it focuses on autonomous restructuring within a locality; it offers the possibility of restructuring the whole of the industry within the locality; while the funds necessary for resourcing the institutions described in (a) are modest compared with the total capital involved in the sector, thus achieving high gearing of public funds.

For brevity I will assume the validity of the analysis underlying the strategy and the practicability of putting it into operation. If this were done, what would be the problems with the class relations involved, in relation to the stated aims of left LEP? (For a particular case see GLC 1985b).

An industrial district of the type proposed would probably be an advance over most urban consumer goods industries in Britain in terms of quality and quantity of jobs(16). The form of the labour market provides considerable bargaining power for the unions, particularly if the union is able to monopolise the supply of skilled labour to firms. But there are important problems, which derive ultimately from the way in which labour is strong in such industrial districts. This strength is based essentially on the economic/social differentiation of markets, particularly the market for the final products and the market for labour power, rather on collective organisation as such. Labour's strength is therefore dependent on the vicissitudes and social structures of these markets. Thus -

(i) in the market for labour power: To use capital's needs for skill as a major source of labour's strength, it is necessary to maintain differentiations within the labour market, both on the supply and demand sides, within production and among workers. Skilled and non-skilled jobs have to be sharply differentiated, something which inevitably weakens collective organisation between 'skilled' and 'unskilled' workers so defined. On the supply side, access to the skilled jobs may be controlled by the union. Given that a rationing system is in operation, it is not coincidental that this has usually been operated on the basis of existing divisions among workers, particularly of gender and race/nationality. Many of the best known 'progressive' industrial districts in the UK and the USA are either overwhelmingly male and white (e.g. shipbuilding, steel fabrication in the UK, construction in New York, printing in London), or have a very sharp sexual division of labour (the New York garment industry). Alternatively, access to skilled jobs may be controlled 'naturally' by social factors: by age, gender, race or nationality. The institution of the family is then central to the structuring of this labour market, whether in terms of position within the family (gender), generation (age), or the family that you are born into (race/nationality) (Solinas, 1982).

(ii) In the market in the final product: if the strategy for good wages and conditions rests on semi-monopoly prices and profits, arising from differentiation of the product, wages and conditions must inevitably be sharply dependent on differences in the success of enterprises in establishing and maintaining semi-monopolistic positions in the market. In 'progressive' industrial districts, there is thus typically fragmented collective bargaining and a wide range of wages between firms in the district. In firms which carry out more routine functions or types of work, conditions will be inferior (Solinas 1982). Moreover, stability of employment is tied closely to the fortunes of the company, which, being fashion-led, are necessarily flighty. There thus tend to be large fluctuations in employment in each enterprise. Even in times of low unemployment, this disrupts collective organisation.

These problems are compounded by other typical features of these industrial districts:

(iii) The unevenness between firms and between different parts of the workforce typically take a geographical form, which compound the difficulties for labour in overcoming these differences. In the 'Third Italy', for example, low unemployment and good conditions (and thus paternalistic capital-labour relations) are maintained in some districts by contracting out low value added work and work at peak times to other localities. (This

relationship has been described as one between
'metropolis' and 'colony' (Brusco 1982, p. 171).
Indeed, the strategy of new industrial districts
involves deliberately erecting barriers between
the district and enterprises in other locations,
especially the Third World; particularly because
of the centrality of skills distinctions (see (i)),
this involves also creating and maintaining a
distinction between metropolitan and non-
metropolitan workers. The strategy is thus inherently, and centrally, localist.

(iv) The fragmented structure of production
and of ownership exacerbates the problems of differentiation. This fragmentation increases the
difficulties of trade union organisation. Moreover,
the usual problems posed by small firms for labour
organisation are in some ways exacerbated by the
particular features of the competitive strategy
typical of the new industrial district. The quasi-
craft identification of the worker with the product
can be used to reinforce paternalistic management,
while the large skills differentials can be used
to create a hierarchy of core and marginal workers
within the firm. Particularly within the smallest
firms, the frequent use of family . labour means
that the power relations within and between families
directly structure the capital-labour relation.

To raise these problems is not to decry some
of the positive elements on which they are built
- the high proportion of skilled work, the integrated nature of much of the work, the identification of the worker with the quality of the product. Nor is it to uphold the standarised product
as 'more moral' than the fashion-differentiated
one. Rather it is to point to the need to develop
industrial structures which do not use non-Fordist
markets and production to weaken collective labour
organisation and solidarity.

One element of an alternative approach must
be to detach as far as possible the conditions
and security of production jobs from the fortunes
of product designs. The necessary differentiation
of design and its inherent unpredictability in
the market would then not need to be associated
with high fragmentation of production and ownership
nor with large fluctuations of employment in enterprises. One 'model' for this is of large production
units using designs from large design departments
as well as outside designers and designer firms.
But to achieve this requires forms of public intervention which go beyond the modulation of markets.

190

CLASS RELATIONS AND LOCAL ECONOMIC PLANNING

Moreover, the instability of contracts for producers cannot be overcome simply through greater concentration of ownership and production, but also requires control over the contracting policies of the retailers/distributors. Within the producing enterprises themselves, the divisions within labour created by capital cannot be overcome simply by public and trade union intervention into labour markets, but requires direct control over management practices. In short, a public intervention to sponsor 'flexible production' requires more direct forms of intervention than the encouragement of progressive industrial district practices if a reversion to localism is to be avoided.

7. Sectoral Versus One-off Investment

I noted above that the structuring of LEP by sector may enable it to correspond to the organisation of workers. But there are ways in which this form of LEP militates against active and direct involvement of workers in the development of policy.

In the first place this appears as a matter of scale; the local sector typically involves thousands of workers; it is practically difficult for the local state to relate to them except via union officials. But in a sense the sector is 'large' only in relation to the scale of the local state's intervention; it is the latter that is the problem. Workers will involve themselves in LEP, and see it as an instrument they can use, only to the extent that it can offer a realistic chance of affecting their jobs. Involvement will not be forthcoming, or maintained, if the state, through shortage of resources, has to turn down the great majority of proposals workers make to it. If this is the case, the primary problems for the state - choosing between sectors, subsectors and firms in which not to intervene - are of only marginal importance to the workforce that is spanned by these choices. The state appears here as the policeman of a rationing system, with the negative control type of functioning that this implies. It will be difficult to involve trade unionists other than at the level of officials, whose specific role is the negotiation of compromises rather than the origination of demands.

The essence of the problem here appears to be the restricted resources available to local government for economic planning. Can this be avoided by a more modest strategy, of reacting to workers' demands rather than actively seeking investments on a planned, sectoral basis? Because workers face their problems in the first place on a plant by plant basis (if not shop by shop), these demands have typically been from workers of a particular plant, especially plants faced with closure. For the state to react to demands to save threatened workplaces would not necessarily be to simply engage in the competitive zero sum game; it could be used to prevent strong labour organisation being broken up by closure or severe restructuring. Similarly, demands of oppressed sections of the workforce are often articulated and organised around at the workplace level in the first place; local government can work more closely with these workers to back up their demands at the enterprise level rather than at the sectoral level. The local state would clearly still need to be selective; but it would be responding to demands from workers where these are being most clearly and forcefully articulated.

An initial point is that this argument is premised on a particular pattern of workers' demands, corresponding to a particular form of labour organisation and consciousness. And the latter is not static: as the economic crisis has deepened over the last 15-odd years, workers have begun to see even their immediate economic problems as soluble not within the plant but at some higher level of the economy. It is true, though, that organisation has changed more slowly than consciousness in this respect. The question is then the degree to which - given the problems of sector planning discussed above - the local state can move ahead of the demands raised by labour.

A reactive policy also faces enormous problems of competitiveness: plants threatened by rationalisation or closure are not necessarily unprofitable, but they are typically technically backward. There is therefore a great risk of the state being seen as a dead hand when its intervention is followed by financial difficulties (given that the role of the previous period of run down or active disinvestment by capital is a less evident cause of these difficulties). Moreover, the acuteness of the problems of competitivity tend to both exclude worker involvement and rule out alternative work arrangements. Decisions have to be taken in a hurry. To the extent that medium-term plans are developed for the enterprise, they may have to be scrapped at short notice to meet financial crises. (The active planning of enterprises, to the extent that this is possible within a market system, tends to be easier the larger the profit margins (Friedman 1977, pp. 24-26). In these circumstances, the workforce may not only be willing to make concessions on staffing, pay, pace of work, etc. (as in any similar firm), but to make these more readily because the enterprise 'is theirs' or is being run 'on their behalf'. The state involvement, instead of shielding the market pressures, may transmit them more forcibly. Whether or not this happens depends on the traditions of the workforce and the way in which the local state becomes involved.

It is true that any local investment strategy, given the present order of resources and powers available to LEP, has to deal with the problem of backward plants in some form. In the majority of cases the local state or enterprise board will not be able to take complete control of the enterprise into which it is intervening (whether as part of a sector strategy or not); it therefore has to have the cooperation of the existing owners. Given that a component of the intervention is to strengthen the hand of labour within the enterprise, the owners will in general only look to the local state for finance if other sources fail - which means that the firm is facing some difficulties. The problems just mentioned therefore tend to arise frequently. But a policy of reacting to demands, rather than seeking out investments which can strengthen labour, exacerbates these problems.

From this point of view, the local state has both too little and too much power. It has the power to save some individual workplaces from closure (and can therefore be expected by labour to try to do so); but to the extent that it lacks any substantial overall control of the sector, this intervention is not qualitatively different from that of an individual firm (and therefore cannot exemplify a method of organising the economy). A sector based approach can perhaps reduce the intensity of this particular contradiction.

Conclusion

I have tried to elucidate some of the problems for a LEP investment strategy that aims to promote organised labour as an economic subject and to use the local state in its support. A part of the argument has been that, while there are tremendous barriers to achieving this, it is precisely the aim of changing such social relations that should be at the heard of LEP, given that net employment creation is beyond its grasp (for a similar argument on GLEB and Women's employment, see Bruegel 1985). The barriers lie partly in the relation of any state structure to an economy which is predominantly privately owned. But there are two other problems which run through this paper which should be highlighted.

The first is the relation between local and national state intervention. We have seen how LEP is very far from being innocent of issues and strategies which have been put forward at a national level. But attempts at local implementation of these strategies, of their exemplification 'in a small way', are often deflected from their aim by the limits of the powers of the local state, or by the very fact of the localness of the inter- vention. This must make one cautious about accept- ing the increasingly canvassed notion of local and national strategies operating in tandem as a division of labour, with local strategies 'pro- viding the level of detail' and 'enabling grass roots involvement' which would otherwise be lacking in national intervention (Boddy 1984; Labour Party 1985). The 'local' is more than a matter of scale and ease of communication; it can also be an element in the generation of particular class politics. To the extent that the latter subor- dinates the demands of labour to those of capital, it precisely militates against grass roots involve- ment.

But the tendency of left local investment strategies to fall into 'localism' lies not only in their implementation in a locality or via the local state, but in the extent to which they fail to make central a consideration of class relations. The issues here are not only of the strategy in the abstract, but also of practice, of implemen- tation. As we have seen, the real content of an intervention can be highly dependent on how the local state positions itself in relation to workers' organisation. The structures and dynamic of current labour organisation, which I have only touched on in this paper, are obviously a crucial determi- nant of positioning. But from the point of view of the local state, relating in the most dynamic way to the self-organisation of labour is the most essential, if the most difficult task.

NOTES

1.   This approach to ideology is developed by Therborn (1980).

2.   There is, however, a specific sense in which the development of LEP within the state has had a degree of autonomy: the immediate pressure for, and conceptualisation of left LEP has come largely from within the Labour Party rather than the trade unions.

3.   The theme of class unity around national competitiveness has been somewhat muted, partly because of the government's different tactics over collaboration with parts of the trade union leadership, partly because of ambiguity in the government's attitude towards capital located in Britain, British capital and foreign capital potentially locating in Britain (cf the remarks above concerning 'local capital' and mobile capital). Conversely, the theme of labour's interests being realised through its subordination to capital has been accentuated.   Some monetarist commentators, for example Samuel Brittan, have explicitly counterposed the latter to the former theme and have argued against any policy of 'national competitiveness'; the government's policy, however, has been far more contradictory.

4.   I have not seen this view propounded systematically, but it is used frequently in discussions of LEP.

5.   This view is very widely propounded both in academic discourse and by political currents ranging from the soft right to the far left.

6.   The question of craft skills in relation to market competition is discussed in section 6.

7.   As with the argument around managerial competence, this can be posed either as LEP reinforcing or substituting for a national strategy, or as LEP fulfilling an intrinsically local function.   This is in fact a quite general bifurcation, and perhaps schizophrenia, of LEP.

8.   Some recent versions of this policy argue that it is not money capital as such whose perogatives need limiting, but merely the existing short sighted or incompetent managers of that capital (the managerial argument applied to the City).   State intervention into the control of money capital is argued to benefit not only productive industry but also the expansion of money capital and its owners, 'savers'.   See Greater

London Council (forthcoming).

9. Looking at this question strictly, i.e. at the international level, none of the 'job creation' mechanisms discussed above addresses itself to the roots of the international stagnation of the capitalist economy, being directed, rather, to altering the conditions of competition for particular parts of capital within that stagnation. A genuine job creation policy would need to at least contribute to a resolution of the underlying crisis. In my opinion, necessary elements of this are either the classic market mechanisms mentioned in (ii) above, and therefore severe attacks on labour, or a planned economy.

10. A more familiar example of the same type of process of 'deflection' is that of the Enterprise Zones. It is now widely recognised that the government's aim of exemplifying the beneficial effects of tax cuts through the Enterprise Zones is vitiated by fact that the tax benefits are local, so that landowners are able largely to recoup them through rises in ground rent. The supposed aim of the policy is 'deflected' and the desired connection between local and national policy invalidated.

11. Unless (a) the local state can achieve more or less complete and permanent control of a sector which (b) is insulated from competition from outside the locality. At present this means small, narrowly defined service sectors.

12. This order of priorities is often evident in process by which left sector strategies have been developed, where the starting point for work on the strategy has been the question of competitiveness.

13. The definition of 'sector' is always dependent on the aim of the sectoral intervention. On both the choice and definition of sectors, note that with a localist policy, intervention is made only into sectors where there is competition from outside the locality, whereas if progressive change in social relations is the focus intervention may be into any sector, including consumer services for example. The range of sectors considered for intervention speaks of the social relations of a local authority's approach.

14. Most of the strategies for private sector dominated industries in Greater London Council (1985a) take the approach, with the partial exceptions of the food and printing industries.

15. That there are, indeed, different overall strategies available for a particular industry

within a capitalist framework is obvious if one looks at the different strategies adopted within particular industries in different capitalist countries and, in some cases, even in different regions within the same country.

16. The 'Third Italy', which is often posed as a model of such an industrial structure, has had one of the fastest growth rates in Europe over the last 15 years.

REFERENCES

Best, M. (1984) 'Strategic planning and industrial renewal: principles to guide the Greater London Enterprise Board's selection of sectors and firms', unpublished.

Boddy, M. (1984) 'Local economic and employment strategies' in M. Boddy and C. Fudge (eds.) Local Socialism, Macmillan, London.

Braverman, H. (1974) Labour and Monopoly Capital, Monthly Review Press, New York.

Bruegel, I. (1985) 'Will women lose out again?', New Socialist, July, pp. 15-18.

Brusco, S. (1982) 'The Emilian model: productive decentralisation and social integration', Cambridge Journal of Economics, 6, pp. 167-184.

Friedman, A. (1977) Industry and Labour, Macmillan, London.

Gamble, A. (1981) Britain in Decline, Macmillan, London.

Glyn, A. and Harrison, J. (1980) The British Economic Disaster, Pluto, London.

Greater London Council, (1983) Small Firms and the London Industrial Strategy, Economic Policy Group Strategy Document No. 4.

Greater London Council, (1985a) The London Industrial Strategy, GLC, London.

Greater London Council, (1985b) Strategy for the London Clothing Industry: a debate, Economic Policy Group Strategy Document No. 39.

Greater London Council (forthcoming) The London Financial Strategy, GLC, London.

Labour Party (1985) Jobs and Industry Campaign, Labour Party, London.

Local Government Studies (1981) Special issue on local economic policy, 7, No.4.

Mandel, E. (1978) Late Capitalism, New Left Books, London.

Minns, R. (1982) Take Over the City, Pluto, London.

Minns, R. (1983) 'Pension funds: an alternative

view', Capital and Class, 20, pp. 104-116.

Murray, R. (1983a) 'The third industrial revolution and the restructuring of the London economy: notes towards a London Industrial Strategy', unpublished.

Murray, R. (1983b) 'Pension funds and local authority investments', Capital and Class, 20, pp. 89-103.

Piore, M.J. and Sable, C.F. (1984) The Second Industrial Divide, Basic Books, New York.

Sable, C.F. and Zeitlin, J. (1985) 'Historical alternatives to mass production', Past and Present, 108, pp. 133-176.

Solinas, G. (1982) 'Labour market segmentation and workers' careers: the case of the Italian knitwear industry', Cambridge Journal of Economics, 6, pp. 331-352.

Stout, D.K. (1979) 'Deindustrialisation and industrial policy', in F. Blackaby (ed.), De-industrialisation, Heinemann, London.

Therborn, G. (1980) The Ideology of Power and the Power of Ideology, Verso, London.

Young, K. and Mason, C. (eds.) (1983) Urban Economic Development, Macmillan, London.

Zeitlin, J. (1985) 'Markets, technology and collective services: a strategy for local government intervention in the London clothing industry', in GLC, Strategy for the London Clothing Industry.

CHAPTER EIGHT

SOCIAL SPACE AND THE PROVISION OF PUBLIC SERVICES:
SEGREGATION IN THE ILE DE FRANCE REGION

MONIQUE PINÇON-CHARLOT

The analysis of social practices should not neglect
the conditions which give rise to these practices.
Such practices, which are embedded in the provision
of public services (schools, museums, hospitals,
transport), can be understood, to some extent,
in terms of the 'objective' characteristics of
these facilities. The research, which we have
undertaken, has enabled us to compare within the
Ile de France Region the distribution of such
services with that of different social groups.
However, this comparison has not been carried out
through an examination of a series of individual
services, for example, the distribution of lycées
with higher level management and professionals
(cadres supérieurs), but by means of comparing
specific clusters of services and the related social
composition of the population. Thus for each urban
commune in the region and for each quartier in
central Paris, the analysis was based on the overall
configuration of services and their characteristics
and on the proportion of different socio-
professional groups amongst the inhabitants of
the various communes and quartiers. In such a
way the types of services were related to the composition of the population(1).
    If the analysis of these social practices
needs to take into account the environment in which
they occur, this does not mean that there exists
some automatic link between the characteristics
of the local environment and these practices. Even
less so, does this environment have a similar impact
on individuals from different social classes and
backgrounds. Yet, while practices cannot be
directly deduced from the specific environmental
conditions in which they occur, it is, none the
less, true that the characteristics of the places

in which individuals live are one aspect of the complexity of factors that determine their social practices.

The first section is concerned with the ways in which the spatial configuration of services is constructed. In the second section we shall show that these services are not randomly distributed in the different social spaces, and that none of the theoretical models commonly involved in the social sciences in France can satisfactorily explain the patterns that have emerged in our research. In order to understand sociologically the spatial articulation between, on the one hand, the structures in the provision of services, and, on the other, social structures, we shall have to clarify the nature of the relationship between the two types of factors which determine social practices. The first category refers to those which are independent of space; the second are those dependent on the spatial context in which the social practice is situated. Unless we clarify these relationships, we shall not be able to say anything relevant about the question of social inequality in access to public services. This latter issue will be the theme of the third section.

The Construction of Public Service Patterns

Our survey of residential areas in the largest urban area of France has enabled us to acquire information, even for smaller units such as the commune and quartier, of the existing 'supply' of public services and certain retailing services which are essential for the resident population. In the course of our survey of the various bodies which managed these services, we have collected a set of almost 400 indicators, which accurately describe the level of local services for 468 urban communes in the region (number counted as urban in 1968 by the National Institute of Statistics and Economic Studies - INSEE). To these communes we have added the 80 quartiers of Paris itself (see Chantrein et al 1976-1977). These several hundred indicators characterise extremely diverse services: sports facilities (Pinçon-Charlot and Rendu 1984), socio-educational, cultural, crèches and facilities linked to education (Pinçon-Charlot 1979), health, work (centres for the professional training of adults, national employment agencies) to communications (Post and Telecommunications, transport), green spaces, commercial facilities and services related to security (police stations and police). These indicators have been <u>constructed</u> on the basis of two main rationales. The first is that of the different administrative sectors which have produced the initial data according to their own institutional logic, while the second results from the researchers themselves who had to make certain decisions about how to utilise the data.

Let us take the example of cultural services. The indicators that we have used correspond to the dominant definition of culture and to the most accepted and established models that regulate how these activities are practised. We are principally concerned with practices (e.g. reading a book, seeing a film or play, looking at pictures or statues, listening to a concert or recital) which aim to appropriate artistic and literary works. Therefore, while measuring the local conditions of access to libraries, cinemas, theatres, diverse entertainments, musical academies and such like, it is important to remember that our attempt to present objectively cultural space, reflects the point of view of the dominant class and, that the construction (no doubt impossible using existing sources - a situation which is revealing in itself) of indicators based on another definition of culture would have led to different results. If we had referred to the anthropological meaning, 'culture' would be extended to incorporate the idea of life style. This would have led us to include other forms of entertainment and activities, such as, bowling alleys, fun fairs, majorette parades, beauty competitions, dances and river-side cafés, motorcross and the PMU (Tote). Hence a definition from the perspective of the dominated classes would have altered the choice of indicators by taking into account elements that were more or less related to working class lifestyles. It is, therefore, essential to have a preliminary definition of culture, otherwise one is liable to commit the mistake of taking a particular 'built cultural space' for <u>the</u> Cultural Space. This argument can, of course, be applied in exactly the same way to sport, health and educational services.

To reduce the complexity of the matrix, constituted by the 400 or so indicators and 548 urban units, we have turned to methods which allow us to analyse large quantities of data simultaneously. We carried out a principal components analysis from which a typology was produced. The 17

categories of communes identified reveal strong
similarities in their types of public services.
With reference to these types, we analysed the
social composition of the resident populations
of the communes and quartiers. In other words,
we compared public service provision with actual
social space. This enabled us to conclude that
the specific characteristics of the services,
brought to light in the typology, corresponded
to significant variations in the social structure
of the communes.

Public Policy and the Production of a Socially Stratified Space

The spatial analysis of the distribution of public
services and social groups (socio-professional
and demographic groups) has led us to conduct an
analysis of social segregation. What was important
was not the relative weight of each social group
in a particular commune, but the system of spatial
relations between social groups within each of
the areas of public services.

Segregation is, above all, relative for there
do not exist in the Ile de France Region any
residential areas that are either totally bourgeois
or totally working class, at least at the scale
of the administrative and statistical units examined
here (commune and quartier). It is only in smaller
residential units that highly segregated populations
occur. Thus the analysis of social segregation
in terms of public services requires a subtle
handling of data which would entail considerable
elaboration, diagrams and statistical tables
(Pinçon-Charlot et al 1985) . In this chapter,
we shall limit ourselves to pointing out the main
conclusions of this research, and in doing so,
run the risk of caricaturing them slightly.

From the outset it is essential to realise
that urban communes and quartiers do not form a
mosiac of types of public service combinations.
The latter are clearly grouped in concentric zones,
which reveal a steady decline in the density and
quality of provision of services from the centre
towards the periphery. From the central quartiers
of the capital to the small communes of the outlying
suburbs, public services diminish in quantity and
quality, reflecting a reduction in income and level
of education of the population and a gradual increase in the number of persons per household.

It is tempting from these initial results

to produce an analysis in terms of social segregation. Yet the typology of communes we have developed only describes inequality in the spatial distribution of public services, albeit an inequality that similarly characterises the spatial distribution of socio-professional groups. Certainly this is the dominant result, but it is nevertheless true that, in the same way that social segregation is, in fact, relative and complex, so the typology does not simply distinguish a dichotomy of public service provision which is inexorably correlated with socio-spatial structures. If this were so, working class groups would also be the furthest away from services and the bourgeoisis and intellectual strata would be systematically the closest to them.

These comments apply particularly to the communes of the inner suburbs. For a few public services these communes are better-off than the Parisian _quartiers_. For example, in the field of sport, they have a wide range of facilities, and, in the sphere of culture, these communes occupy a privileged position (admittedly not when it comes to major cultural centres but more for the public libraries, municipal music schools and proximity to technical colleges). In other services, such as crèches, the working class communes in the inner suburbs have the most places. Within the inner suburbs, that is at an equal distance from the centre of the capital, public services vary according to the social structure of the resident population (Pinçon-Charlot and Rendu 1982).

Public services are not randomly distributed in the different social spaces. In the Paris region, those classes which enjoy the best conditions of residential life, due to their privileged location in the central areas of the capital are the bourgeoisie (major industrialists and merchants), followed by the traditional petit bourgeoisie, intellectual strata and those involved in personnel services (tied to the upkeep of the well-to-do classes, to office cleaning and employment in hotels). Conversely, the working classes are heavily over-represented in the under-equipped communes on the periphery of the region. Middle management and employees are fairly uniformly present in all types of areas.

However, it is advisable to superimpose upon these social inequalities and their associated differences in public service provision, those inequalities engendered by differences in the structure of provision. Equidistant from the centre, variations, which cannot be reduced to quantitative terms, have been observed. One can identify clusters of public services which reach a relatively high density or quality in the most working class areas and different combinations of services characteristic of localities where the middle classes are over-represented.

Over-emphasis on inequalities between the distribution of services and their relationships to the social composition of different areas of the region carries the risk of reducing the social effects of urban service politics to the measurement of the physical distances which separate members of each social class from the public services necessary for their existence. We have therefore noted the connections between 'objective' physical distance which separates agents and services (focusing on certain social categories and types of collective consumption), and the social distance between the same agents and activities (the latter being subjective, 'lived' distance). In fact, spatial distance is never reducible to an objective measure, even when taking into account travel time. This distance is always experienced in terms of the relationships of different social groups to space and to their potential use of public facilities. Thus, the ease of travelling long distances has a tendency to increase with income and education levels. In effect, the inclination to carry out a particular activity has the effect of reducing the subjectively perceived (and lived) distance (Pinçon Charlot and Rendu 1982).

From Theoretical Models to an Interpretation of the Results

We are faced with the paradox that none of the theories commonly suggested as explanations of variations in the provision of facilities is able to explain satisfactorily the previously described situation. As we shall show, each of the theories is only able to offer a partial explanation.

For example, the model of the operation of the market tends to see the relationship between urban structures and social structures as the result of a spontaneous adjustment between supply and demand, but this only helps us to understand the distribution within the private sector. Those facilities provided by the private sector are associated with the wealthiest sections of the population. However, this model encounters difficulties when it comes to understanding the distribution of such public facilities as major theatres, which have been decentralised to the working class suburbs of Paris. All 'marketing' studies have shown there is hardly any demand for this type of provision in these suburbs.

A second model is that based exclusively on the thesis of social control and discipline (normalisation) of the working classes (Fourquet 1973; Foucault 1975; Dreyfus 1976; Guillaume 1976) through the provision of public services(3). This can quite well account for the provision of certain types of public services. As an example of this type of facility one can think of sports clubs whose managers sometimes explicitly make demands for their creation on the basis of the ability of these clubs to prevent juvenile delinquency. Yet this model is totally incapable of offering a coherent explanation of the spatial proximity of the middle and upper classes to the provision of many facilities, especially those related to culture and education. To the extent that it is the middle and upper classes who make the greatest use of these facilities, it is simply not possible to reduce the presence of sports clubs to the function of establishing an environment for the working class. Even less so, does the density of police and police call boxes, of which there are greater number in wealthier areas than in working class areas, substantiate the theory of social control.

Finally, in certain areas, the quality and quantity of facilities seem to be unevenly correlated to what one supposes are the 'needs' of the resident population. We have shown that the spatial pattern of decreasing provision of facilities from the centre of the capital to the periphery of the urban region does not always occur; it is broken in places, at least for certain types of collective consumption, by a provision which is related more to class structure than distance from the centre of Paris. A highly working class area in the inner suburban ring may give its inhabitants fuller or better services in some domains in comparison with the two other types of areas in the same zone, but provide worse services in other domains.

Our research has also brought out a correlation between the location of cultural facilities that inculcate certain cultural practices and establishments (municipal conservatoires and public libraries) and the location of a section of the working class. There is also a correlation between the most expensive cultural facilities, such as museums, cinemas and especially theatres, with social groups that are already endowed with a certain cultural knowledge and disposition which enables and encourages them to use these services. This is the case of many teachers who, despite their modest incomes, tend to live in areas which are the best equipped with these types of cultural facilities.

One might think that the correlation between groups with cultural capital and provision of services would lead one back to the concept of the reproduction of the labour force as the principal explanation of the patterns we have noted (see Preteceille et al 1975). However, we should remember that the correlation between the distribution of facilities and of social groups is far from consistent. For example, there is under-provision in many working class housing estates, and this does not seem to support the idea that there is a correspondence between provision and 'the need to reproduce the labour force'. At the same time, the notion that 'needs' are related to the reproduction of the labour force poses problems as well. Above all, does not the provision of cultural facilities represent the image that certain local politicians have of 'working class' cultural needs and/or the result of competition in the political field?

Consequently none of these theories completely account for the complex reality of the collective consumption. It is as if, in certain cases, the logic of the market place, social control or reproduction of the labour force was the only one in operation. Yet, as we have previously demonstrated in our initial analysis of the specific combination of services, in most social environments there are a number of factors which intervene at the same time. Proposals for the production, management and utilisation of services, in effect, depend on a number of social agents occupying different, and sometimes, contradictory, social roles. Thus the multiplicity of factors involved leads us to specify a set of conditions which can rarely be analysed in terms of a single logic. Although theories of social control, market forces and reproduction of the labour force do account for some of the provision of services, they in no way totally exhaust all the relevant factors.

In examining the social consequences of the spatial distribution of public services identified in our study, we have invoked a problematic founded on the idea that any particular practice follows from the interaction between the habitus and the conditions in which that practice occurs. The habitus is defined by Bourdieu (1977, pp. 77-78) as:

> the strategy generating principle enabling agents to cope with unforeseen and ever-changing situations... a system of lasting transposable dispositions which, integrating past experiences, functions at every moment as a matrix of perceptions, appreciations and actions and makes possible the achievement of infinitely diversified tasks.

The habitus is therefore a logic derived from a common set of material conditions that is internalised and operationalised by an individual or group in the course of its experience(4). Without any doubt the position occupied by the individual in the relations of production, or by members of the family, if the individual does not work, gives the acquired experience one of its essential dimensions. This obviously extends beyond the sphere of work in its strict sense and covers numerous aspects of his/her lifestyle, so that it is not unreasonable to speak of a class habitus to designate the totality of common aspects inscribed in the habitus of members of the same social class through the similarity of their experience. As for the circumstances in which the practice occurs, these bring together the external conditions engendering the practice. These include not only the physical or material circumstances (in the case of services, this covers principally proximity, capacity, times of opening, use and cost), but also the social conditions in which the practice would be engaged in.

Included amongst the social conditions is the social mix of different social categories at the local level which is a particularly important consideration. The relative proportion of different groups in a specific locality and the relations of domination formed amongst them contributes a contextual variable in defining the conditions underlying social practices. One can put forward the hypothesis that the likelihood of members of different classes making use of services depends partly on the population which uses them, the intensity and style of usage and the social image of the area, all these factors being linked to the social composition of the local residential environment.

It is thus vital not to separate the analysis of practices into the study of the characteristics of the services and local social conditions. Both aspects contribute to define the social conditions in which the practices are produced. This approach breaks with the functionalist aspects of the models we have criticised, in that the social impact of a combination of services cannot be read off mechanically from the nature of the services themselves, for the latter also depend on the social context in which they are likely to have an effect. This is why we have examined the relationship between the objectively measurable physical distance, which separates social agents and services, and the social distance between these agents and the relevant practices, with distance being experienced subjectively. As we have argued earlier, spatial distance cannot be reduced to an objective measure even when transport time is taken into account. The same physical distance can thus have a differential effect on the intensity of use of a public service by different social categories as a result of the varying mobility of these groups. Furthermore, physical and social distance can reinforce each other to the extent that their interaction may encourage or discourage groups from using services. On the other hand, these two types of distance may work in the opposite way and partially cancel out each other.

Let us take the example of those groups with the least amount of educational and cultural capital. If it appears self-evident that physical distance will reinforce social distance in discouraging these groups from engaging in certain social practices, it is, on the other hand, not so easy to predict what the effect of diminishing spatial distance would be on their use of services. A priori one might assume that a decrease in physical distance might help to overcome the handicap resulting from social distance. In any case this is the strategy followed by those in favour of cultural action. This can be defined as a social experiment in which deliberate intervention upon the conditions of practice has the explicit aim of influencing the practices themselves. For example, this is the type of cultural strategy pursued by some communist municipalities. However, as we have developed at length in our publications, the reduction of spatial distance, and especially the intervention of professionals does not systematically lead to the widespread diffusion of social and cultural practices. Indeed, in some cases this intervention may paradoxically strengthen the influence of social distance, and even more radically still, separate the working classes from practices and places which are not theirs.

Future Research Directions

In the course of our research we have gained knowledge of the socio-spatial distribution of public services in the largest urban region in France and have shown that spatial proximity to services for the working class represents a favourable, perhaps necessary, though not sufficient, condition for overcoming differences in access to services (Pinçon-Charlot and Rendu 1982). We have also highlighted the social, economic and cultural processes which result in the omnipresence of the intellectual fractions of the salaried middle classes in the domain of collective consumption (Pinçon-Charlot et al, 1983; Pinçon-Charlot and Garnier 1985). Our current research is concerned with outlining certain aspects of decision-making in the provision of services through a study of high-ranking civil servants. Our hypothesis is that the discourse of these civil servants, and the position they adopt in relation to these services, cannot be isolated from their own experience and their ideas about this provision. Their ideas are derived from their practices, and from those of their family and class, from the usage or non-usage of a certain type of service and from the social conditions of their utilisation of these services.

Thus in considering both the social context in which facilities operate and the disposition of the agents involved in their provision, we have tried to avoid a mechanistic or functionalist analysis of the social effects of environment. Instead the institutional and social logic put into practice by local and central government civil servants, by the agents involved in the organisation and management of these services and by the local politicians... encounters the logic of those social

categories who utilise or do not utilise these
services. It is in the constant interaction, which
is basically not between individuals but mediated
by the logic of administrative and political
institutions and the system of social stratification, that one can reach an understanding of
the social nature of the provision of services.

NOTES

1. This research project was carried out
in 1973 with Edmond Preteceille and Paul Rendu,
researchers at the Centre de Sociologie Urbaine.
2. Geography, sociology, history and political science have all been concerned with de-
limiting social divisions in space and the different
ways in which space is used. This should result
in interdisciplinary work, but it has so far been
rare. For a recent example see the publication
of the Laboratoire de Science Sociales de l'Ecole
Normale Supérieure (1983).
3. An example of such an interpretation
can be seen in the extract from Fourquet et al,
(1983, pp. 127-128):

> Public services have been historically
> constituted as instruments of domination
> and the territorial settlement of the
> waves of population liberated by the
> destruction of artisanal and agricultural
> family production. These services are
> not there instead of the former extended
> family for the purpose of producing and
> reproducing the modern nuclear family...
> The concept of normality comes to the
> fore in the domain of public services,
> especially medical and educational...
> A normative social class assigns models
> of normality to public services and im-
> poses types of control and standardisation
> that the nuclear family is incapable
> of doing.

4. For a clear outline of Bourdieu's
theoretical and empirical work, especially his
contribution to the sociology of culture, see
Garnham and Williams (1980). This issue of Media,
Culture and Society also contains a bibliography
of Bourdieu's work in English until 1980
(pp. 295-296). Research by Bourdieu and others
at the Ecole des Hautes Etudes en Sciences Sociales
comes out regularly in the Actes de la Recherche
en Sciences Sociales.

REFERENCES

Bourdieu, P. (1977) Outline of a Theory of Practice, Cambridge University Press, Cambridge.

Bourdieu, P. (1980) Le sens pratique, Les Editions de Minuit, Paris.

Bourdieu, P. (1984) Distinction. A Social Critique of the Judgement of Taste, Routledge and Kegan Paul, London.

Chantrein, M., Pincon, M., and Preteceille, E. (1976, 1977) Indicateurs de l'équipement collectifs en région parisienne, 2 vols. Centre de Sociologie Urbaine, Paris.

Dreyfus, J. (1976) La ville disciplinaire, éssai sur l'urbanisme, Editions Galilée, Paris.

Foucault, M. (1977) Discipline and Punish: The Birth of the Prison, Allen Lane, London.

Fourquet, F. et al, (1973) 'Les équipements du pourvoir' Recherches 13, pp. 127-128.

Garnham, N. and Williams, R. (1980) 'Pierre Bourdieu and the sociology of culture: an introduction', Media, Culture and Society, 2, pp.209-223.

Guillaume, M. (1976) Le capital et son double, PUF, Paris.

Laboratoire de Sciences Sociales de l'Ecole Normale (1983) 'Territoire et territorialité', Territoires, 1.

Pinçon, M. (1978) Besoins et habitus. Critique de la notion de besoin et théorie de la pratique, Centre de Sociologie Urbaine, Paris.

Pinçon-Charlot, M. (1979) Espace social et espace culturel. Analayse de la distribution socio-spatiale des équipements collectifs et éducatifs en région parisienne, Thesis, Centre de Sociologie Urbaine, Paris.

Pinçon-Charlot, M. (1983) 'Les couches moyennes salariées, agents et usagers des politiques publiques: de communauté du thème à la cumulativité des resultats' in C. Bidou et al. Les couches moyennes salariées. Mosaique sociologique, Ministère de l'Urbanisme et du Logement, Paris.

Pinçon-Charlot, M. and Garnier, Y. (1985) 'Ségrégation sociale et équipements culturels. Le cas de l'enseignement musical dans un conservatoire municipal' Les Cahiers de Sociologie de l'Art et de la Litterature, 3.

Pinçon-Charlot, M. and Rendu, P. (1982) 'Distance spatiale, distance sociale aux équipements collectifs en Ile de France: des conditions de la pratique aux pratiques', Revue Francaise de Sociologie, 23, pp. 667-696.

Pinçon-Charlot, M. and Rendu, P. (1984) 'Espaces sportifs, pratiques sportives' in Sports et Sociétés contemporaines, pp. 299-304.

Pinçon-Charlot, M., Preteceille, E. and Rendu, P. (1985) Ségrégation urbaine. Classes sociales et équipements collectifs en région parisienne, Editions Anthropos, Paris.

Preteceille, E., Pinçon-Charlot, M. and Rendu, P. (1975) Equipements collectifs, structures urbaines et consommation sociale. Introduction théorique et méthodologique, Centre de Sociologie Urbaine, Paris.

Publications of the Centre de Sociologie Urbaine can be obtained directly from it at 118 rue de la Tombe Issoire, 75014 Paris.

CHAPTER NINE

THE ROLE OF LABOUR AND HOUSING MARKETS IN THE
PRODUCTION OF GEOGRAPHICAL VARIATIONS IN SOCIAL
STRATIFICATION

CHRIS HAMNETT AND BILL RANDOLPH

Introduction

In his 'The Condition of the English Working Class
in 1844', Engels sketched his now well known
description of the spatial distribution of social
classes in Manchester. In doing so he recognised
that social stratification necessarily has a spatial
dimension. Exactly 130 years later, Giddens (1973,
p. 199) pointed to the fact that:

> In the past development of the working class,
> in the Western societies at least, the in-
> fluence of neighbourhood and regional segre-
> gation has been fundamental to class struc-
> turation and class consciousness. Such segre-
> gation has taken various forms. Thus, in all
> the advanced societies, there are regional
> variations in the distribution of workers
> in manual labour, particularly in manu-
> facture...But there have also been, his-
> torically, important divisions between com-
> munities. It is with some truth that the
> archetypical 'proletarian worker', a member
> of a clearly distinct 'working class culture',
> and strongly class conscious, has been associ-
> ated with industries, such as coal mining,
> which have grouped workers together in isolated
> villages or towns.

On the face of it, this might appear to indicate a continuing awareness of the role played
by spatial variations in social stratification.
Nothing could be further from the truth. Although
there has been a long tradition of 'Community
Studies' within sociology (Dennis, Henriques and
Slaughter, 1957; Frankenberg, 1966), the study

of social stratification and the study of spatial variations in social composition tend to have proceeded in splendid isolation. Whereas most urban sociology and urban social geography until recently, tended to take the structures of social stratification as a given and ignore them; sociological grand theory has tended to examine social structures and processes as though they existed on the head of the proverbial pin. As Beshers (1962, p. 41) perceptively pointed out over twenty years ago:

> No locale nexus for social relationships is provided by (the) macroscopic view of stratification...(It) ignores the variations in stratification within society - variations among regions, among subcultures, between rural and urban areas, and even within social strata.

Until relatively recently, the spatial dimension of social processes has been a 'lost world' to many social scientists. Society and space have been analysed as though they existed independently of one another. But, as Castells (1977), Duncan (1979), Santos (1977) and Sayer (1979) have all pointed out, society is necessarily spatially structured and all social structures and processes possess an inescapable spatial dimension. As a result of this theoretical and analytical lacunea a number of recent authors (Giddens, 1979; Urry, 1981) have pointed in general terms to the need to reincorporate the spatial dimension into the study of social processes. As Urry (1981) has argued: "sociology (apart from its urban specialism) has tended to pay insufficient attention to the fact that social practices are spatially patterned, and that these patterns substantially affect these very social practices" (p. 456).

The lost world has once again been rediscovered but, like so many expeditions into the theoretical 'terra incongitae', traces can usually be found of an earlier generation of explorers. Ray Pahl (1970a) was advocating such a position from the mid-1960s onwards in papers such as 'Spatial structure and social structure' in which he tried to outline the principles of a spatial sociology which would seek to come to terms with the fact that: "The interrelationship between the social and the spatial has not been adequately conceptualised in sociological analysis". Pahl (1970b) subsequently attempted to utilise this perspective to construct a distinctively urban sociology which would focus on spatial inequalities in both the provision of and access to scarce urban resources and their mediation by urban managers. We do not intend to get embroiled here in the subsequent debate regarding the validity of a spatial focus as "a defining feature of urban sociology" (Saunders, 1981; Harris, 1983; Kirby, 1983). Instead, we merely wish to reassert the shortcomings of any analysis of social stratification which does not incorporate some consideration of the spatial dimensions of stratification.

Labour markets, housing markets and social stratification

That social stratification has a spatial dimension is undeniable. But, before we consider the ways in which geographical variations in social stratification are produced and constituted in and over space, it is first necessary to attempt to define what we mean by social stratification a little more clearly. The classic distinction is between the marxist definition of social classes in terms of relationship to the ownership and control of the means of production and the Weberian definition of social stratification in terms of position in a variety of markets. This weberian definition of classes should not be confused however with the idea of social status groups which are "based upon the distribution of prestige or social honour rather than, on material economic inequalities" (Saunders, 1980, p. 74).

One of the best known attempts to develop a specifically weberian analysis of class based on market power and position is Rex and Moore's (1967) analysis of housing classes. While they recognised that power in the housing market was in large part a reflection of power in the labour market, they nevertheless argued in their analysis of housing classes that it was possible to occupy one class situation in relation to the means of production and another with relation to the distribution of domestic property. The validity of Rex and Moore's formulation has been subject to considerable criticism, both theoretically and empirically (see Haddon, 1970; Saunders, 1980) and it is not our intention here to get enmeshed in this debate. Suffice to note that Saunders (1984) has recently argued that the notion

of housing classes 'overextends class theory' and
that: 'Class relations are constituted only in
the sphere of production'. He thus argues that
whilst "housing tenure cannot be a factor in class
structuration...(it)...can be an element in social
stratification". As Payne and Payne (1977, p.
134) have put it:

> It is through the mechanism of housing that
> the major life experiences, conventionally
> associated with occupational class, are deter-
> mined; housing's relevance for stratification
> is not just as an index of achieving life
> chances, but as the means by which the in-
> equalities of the occupational structure are
> transferred into the wider social structure.

What Saunders has done, in effect, is to fall
back on a marxist conception of class embedded
within a wider weberian conception of social stratification. This position has the advantage
of acknowledging that class relations are constituted in the sphere of production whilst recognising the major role played by consumption relations in the production of social stratification
as a whole. As Saunders (1984, p. 207) points
out:

> 'While consumption is constrained by pro-
> duction, it is not determined by it; processes
> of consumption have their own specificity'.

This is the position that will be adopted in the
remainder of this paper.

This position represents a valuable starting
point but it does not address the central question
to which this chapter is directed, namely, <u>how
does social stratification vary spatially and how
are such spatial variations produced, reproduced,
maintained and changed</u>. Whilst a focus on differences in occupational structure, employment,
economic activity rates and the like must be central
to any analysis of spatial variations in social
stratification, it is clear that these are not
the only dimensions of social stratification. Nor
are differences in labour market structure from
one part of the country to another the sole source
of spatial variations in social stratification.
On the contrary, differences in gender, ethnicity,
religion (in parts of the country) and housing

market position can all be legitimately considered as both dimensions and determinants of stratification.

    In this chapter however, we wish to adopt a more specific focus and examine the role played by the interaction of the labour market and the housing market in the production of spatial variations in social stratification. This is not to suggest that the other dimensions of stratification are not important. They are. It is rather that the housing market (defined as both the structure of housing market opportunities differentiated by tenure, price, dwelling type, quality etc. and the constraints on access) is considered to constitute along with the labour market one of the principal determinants of spatial variations in social stratification at the sub-regional and local level. The principal focus of our attention in the remainder of this chapter is therefore on how the labour and housing markets interact in and over space to produce, maintain and change spatial variations in social stratification at the regional, sub-regional and intra-urban levels. Gender, ethnicity and other dimensions of stratification are only considered explicitly insofar as they enter into the analysis of labour and housing market interactions.

The Production and Maintenance of Local Variations in Social Stratification

Income and wealth play a major part in controlling access to a wide variety of scarce social resources, and as paid employment constitutes the principal source of income for most of the economically active population and their dependents, there can be little doubt that labour market position, occupation and income collectively constitute the principal determinant of social stratification at the national level. But the concept of national social stratification is clearly an aspatial abstraction. Neither the labour market nor the individuals who constitute it exist on the head of a pin. On the contrary, as most people are only too well aware, the labour market is highly spatially differentiated. At any one point in time, there are marked spatial variations in both industrial and occupational structure, and in the demand for and supply of labour of different types. Such differences are clearly crucial in shaping the broad geographical linaments of social stratification but, insofar as they say nothing about where people actually live, they constitute only one determinant of the spatial variations in social stratification from one area to another. The problem, as Pahl (1970a) has recognised, is to relate the occupational structure to the structure of residential social stratification. As he puts it: 'given a specific historical, cultural and political situation, how does the product of a given division of labour adjust and adapt itself in space...'(p.187).

Although the precise wording of Pahl's question can be queried, he rightly identifies the relationships between the labour market and the housing market as the key to understanding the geography of social stratification. One of the most important ways in which 'the product of a given division of labour' is distributed in space is residentially, and the residential distribution of different occupational and income groups depends in part upon both the existing geographical distribution of the housing stock, which will be differentiated in terms of age, tenure, price and quality, and the structure and operation of the housing market which allocates that stock to different groups in the population. As Castells has observed:

> urban stratification and segregation are not the direct projection in space of the system of social stratification but an effect of the distribution of the (social) product among the subjects (i.e. individuals) and of the housing product in space, and of the correspondence between these two sets of distributions (1977, p. 171) (our inclusions).

What Castells is arguing, correctly in our view, is that spatial variations in social stratification are the product of the spatial 'correspondence' or 'non-correspondence' between differences in labour market position and other sources of social stratification, the structure of the housing market in space and the constraints on access to it. The spatial structure of social stratification is just as much a product of where people live as it is of their other social characteristics. Variations in social composition can thus be seen as the outcome of both the structure of industry and employment and the structure of the housing market. The two markets comprise the major structuring mechanisms through which social stratification is constituted in space and in the daily reality of people's lives. Whether, where and at what people work is necessarily related to the question of where people live. Work and residence are intimately and reciprocally inter-related in and over space.

Unfortunately, the implications of this re-ciprocal relationship between work and residence for the production of spatial variations in social stratification have not been generally appreciated. For the most part the two markets have commonly been analytically separated and treated in splendid isolation. Economic or industrial geography has concentrated on where people work and urban social geography on where they live as though the two things were unrelated. Thus, whilst the importance of the housing market in the 'sifting and sorting' of the population in space has now been widely recognised by a variety of writers, the structure, and indeed the very existence, of the population to be sieved through, the housing market is commonly treated as an unproblematic given. It is merely assumed to be already in existence, ready and wait-ing to be processed through the housing market. Nor, for the most part, are spatial variations in the structure of the housing stock explicitly problematised. Like the population, the housing stock is generally assumed to be there, awaiting its quota of households. The fact that the geo-graphy of social stratification varies by residence as well as by work place seemingly goes unappreciated.

Although the last few years has seen a growing awareness of the role played by the spatial division of labour (Massey, 1978, 1983a, 1983b, 1984; Lipietz, 1980; Cooke, 1981, 1982) in the production and maintenance of spatial variations in class structure, most of this literature treats the spatial division of labour as though it takes place wholly independently of the housing market. At best, the housing market is assigned a residual status as an 'also ran' in the production of vari-ations in social stratification. To this extent, it can be argued that the analysis of socio-spatial change has generally suffered from a form of regional-economic determinism which has tended to treat changes in employment structures as the sole determinant of local social stratification. Massey's work on the restructuring of class re-lations both socially and spatially has tended to focus on economic changes at the inter-regional level, and Urry's (1981) discussion of class relations in space also suffers from the fact that it largely ignores questions of housing and residence. In his discussion of 'local social structure' and its impact on 'local social movements', Urry makes only fleeting reference to the role of 'residence' in relation to those factors which help determine the nature of local labour markets. That residential structure is itself an important constitutive element in the formation of local social structure goes seemingly unappreciated.

As a result, the relationships between labour market and housing markets and their role in structuring social stratification in space have generally been neglected. It is rather ironic that the impact of the housing market on labour supply and demand has been given most attention by a number of rather traditional labour market studies (see Bramley, 1979, 1980; Cheshire, 1979; Evans, 1981; Gleve and Palmer, 1981; Metcalf and Richardson, 1980; McGregor 1979; and Evans and Richardson 1981). But, as we shall argue, both housing markets and labour markets have their own particular spatial characteristics which need to be explicitly incorporated into any coherent analysis of variations in social stratification from one area to another.

The Problem of Scale

To say that work and residence are intimately and reciprocally interrelated in and over space is perhaps to state the obvious, but it raises a crucial question regarding the geographical scale of the interactions. Although there are distinct regional differences in industrial and occupational structure, activity rates and levels of unemployment (Gillespie and Owen, 1981; Massey, 1978; Martin, 1982) and in the structure of housing tenure and house prices (Hamnett, 1983a), at the local level both the labour market and the housing market are geographically limited by the constraints imposed by commuting into a series of relatively discrete geographically defined individual markets. While the structure and operation of the two markets cannot be analysed or understood at the local level alone, it is clear that - with the exception of some of the larger metropolitan areas - the two markets possess a high degree of local correspondence in space. They are spatially linked by the socially differentiated time and money constraints on journey to work trips. As Pahl et al (1983, p.81) comment, 'most people expect to find work within an easy journey from home'. Indeed, as Johnson, Salt and Wood (1974) have pointed out, there exists a reciprocal relationship between the two markets:

> From the point of view of the individual worker, the location of his (sic) residence also limits his (sic) choice of job. The locations of job opportunities and places of residence are mutually linked in space by the commuting 'tolerances' of workers. Given the location of work-places in centralised nodes, the area over which housing opportunities will be sought by a household depends on its members' tolerance to commuting. In the same way, given the location of residence, the workers in a household seek employment opportunities over an area defined in terms of commuting tolerance (p. 28).

The result is a complex patchwork of partially overlapping 'local labour market areas' in which "the great majority of people live and work" (Goddard 1982). Goddard argues that the local labour market represents the most appropriate scale at which the analysis of spatial variations in economic development should be focused. There is an analytical problem here, however, as it can be argued that the relative importance of the two markets in the production and maintenance of spatial variations in social stratification varies according to the scale of the analysis. While the labour market can be seen as the predominant influence upon the overall class and occupational structure of a region or metropolitan area, the spatial distribution of the housing stock and the structure and operation of the housing market can be argued to be more important in 'fixing' the social structure of particular local areas. To put it crudely, the labour market can be argued to create the overall class character of a regional or local labour market, whilst the housing market sorts it out spatially within local labour markets. Changes in the employment and occupational structure of local labour markets are translated into spatial variations in social stratification through the operation of a spatially differentiated, socially stratified and institutionally structured housing market (Figure 9).

Figure 9: The Differential Impact of Labour and
Housing Markets in Structuring Spatial
Variations in Social Stratification

Level of Spatial Resolution	Predominant Structuring Process	
	Labour Market	Housing Market
National Regional Sub-Regional Local Neighbourhood	↑	↓

This argument oversimplifies what is, in
reality, a more complex set of reciprocal deter-
minations. Just as there exist significant local
variations in industrial structure and the structure
of labour demand which can and do exert a major
influence on the social and occupational structure
of local areas, the structure of housing market
opportunities can also be argued to play an increasingly important role in influencing the changing spatial location of certain types of industry.
Not only is the structure of local housing opportunities an important consideration for some
companies seeking to relocate or establish production, but geographical variations in the housing
market can also be seen to embody different pools
of labour with different skills and attractions.
In other words, the local structure of housing
opportunities can and does have an 'impact back'
on the structure of employment.

The Changing Form of the Labour-Housing Relationship

We have already stressed that access to the housing
market is constrained by income and position in
the labour market. But the relationship between
the labour market and the housing market is not
a simple one-way determined one. While position
in the housing market is partially a product of
position in the labour market, it cannot simply
be 'read off' from it. So too, spatial variations
in the structure of the housing market are not
simply the product of spatial variations in the
structure of the labour market. Although it can

be argued that the form, nature and location of housing provision over the last 150 years or more has tended in broad terms to reflect the industrial, occupational and income characteristics of local labour markets, the reflection is by no means a perfect one and, with the exception of some of the nineteenth century integrated industrial and residential developments such as Port Sunlight and Bourneville, there is no simple correspondence between the two markets either nationally or locally. There are links to be sure, but in general both markets possess their own independent logic, structure and history, and both operate relatively autonomously. We therefore need to incorporate an understanding of the changing employment structure of different areas on the one hand, with an understanding of both the production and consumption of housing on the other it we are to understand how and why the local structure of social stratification varies from place to place, area to area, and region to region.

To argue this is not merely to argue, after Saunders (1984) that the logical primacy of production does not, of itself, demonstrate its social determinacy, and that consumption processes possess their own specificity, true though this is. Nor is it just to argue that housing tenure is a major dimension of social stratification irrespective of space. It is also to assert the truism that housing itself must be produced before it can be consumed, and that the historical legacy of spatially differentiated forms of housing production interacts with the structure of locally available employment opportunities to structure differences in residential social stratification.

To summarise our argument so far we can say that while changes in the structure and organisation of the labour market stemming from changes in the labour process will, by changing the skill, status and income characteristics of both the resident and the immigrant labour force, lead to a restructuring of the broad pattern of demand for housing in an area, changes in the structure of the housing market cannot be regarded as a passively determined response to changes in the labour market. On the contrary, the spatial structure of the housing market itself feeds back into the labour market through its influence on the spatial supply of labour. The changing structure of the housing market is, to a considerable extent, an autonomous outcome of changes in the production of housing as a commodity in its own right. Changes in the structure of both the housing and the labour market are autonomously constituted processes. It follows that there is no direct or necessary correspondence between the two markets. Instead, the very real relationships which exist must be viewed as essentially socially, spatially and historically contingent.

Spatial and Temporal Variability in the Two Markets

It is this relative autonomy of the two markets and their contingent interrelation which partly accounts for the variety (both spatial and temporal) of outcomes. In the nineteenth century the great majority of the urban labouring poor were, as Stedman Jones (1971) has convincingly shown, effectively condemned by poor and relatively inaccessible and expensive forms of transport to live as close to their work as possible in densely overcrowded slums. The 'employment linkage', as Vance (1966) has termed it, was such a strong factor in nineteenth century urban social and physical structure that it is easy to jump to the conclusion that the housing market is always necessarily dependent upon the nature of the labour market. But, as Dyos (1968), Kemp (1981), Ball (1981), and Aspinall (1982) have all shown, the nineteenth century housing market possessed its own distinct and independent logic and organisation.

There is also a good case to be made that as developments in transportation have progressively allowed increases in the size of individual labour markets, or 'daily urban systems' as Berry (1970) has termed them, that the importance of the 'employment linkage' has gradually weakened for the more affluent and mobile sections of the community. As a consequence local labour markets have become progressively internally differentiated in terms of both their housing stock and their social characteristics. They have, in other words, become spatially stratified on an increasingly extensive geographical basis. Although the structure of local employment still remains the overriding determinant of the general structure of social stratification by occupation and income at the local labour market level, the increasingly differentiated spatial structure of stratification within these labour markets is predominantly a reflection of the spatial structure of housing market opportunities by price, house type, size, quality and tenure.

While the residential social structure <u>within</u> labour market areas may be argued to have become increasingly differentiated through the spatial structure of the housing market, it is also the case that changes in the spatial organisation of production (Massey 1983b, 1984) are leading to growing differences in the occupational structure <u>between</u> different labour markets. In particular, the concentration of corporate headquarters in a relatively small number of large cities (Noyelle, 1983; Simmie, 1982) has increased the concentration of professional and managerial employment in these cities and has resulted among other things in the 'gentrification' of certain inner city areas (Hamnett 1984c). The growing polarisation in employment structure in these cities into a highly skilled and highly paid elite and a low skilled or deskilled service labour force (Sassen-Kolb 1984) is resulting in an increasingly polarised housing market and residential social structure. Simultaneously, some 'sunbelt' labour markets are undergoing a rapid expansion whilst other 'snow- belt' labour markets are experiencing equally rapid industrial decline. Such changes are having a profound impact on the spatial structure of social stratification.

Spatial and Temporal Fixity

The changes discussed above raise the question of how far the labour and housing markets are dif- ferentially constituted in time and space and with what consequences. As Pahl (1970) has indicated: 'The housing market and the job market need not be congruent, there may be availability of houses but jobs may be declining locally or there may be approximate parity between the total number of jobs and dwellings but the dwellings may be too expensive in relation to local wages' (p.216b). Taking the temporal dimension first, we would argue that once built, housing as a physical entity tends to be relatively more enduring than the employment structures with which it may have been initially associated. As Johnson, Salt and Wood (1974) have pointed out: 'The immobility of housing is an... important characteristic...Since a high proportion of aggregate stock is not of recent construction, its location pattern reflects past distributions of population and economic opportunity' (p.15). While the local employment base of an area can virtually vanish overnight, the links which bind people to their homes are often more permanent and enduring. When a firm gradually shrinks or collapses, the workforce and the houses they occupy do not disappear with it. The houses remain in place and so do the majority of the labour force, particularly the less skilled. This is not a new phenomenon as Charles Booth commented of late nineteenth century London: "Trades leave, people stay" (quoted in Stedman Jones 1971 p.154). Once in place, residential structure responds only slowly to shifts in economic activity.

The fact that housing tends to have a higher degree of 'temporal fixity' than employment has two important consequences. First, and following from the argument above, large sections of the local labour force can effectively be left economically stranded high and dry in areas of employment collapse. Apart from ties of family and community, they are often 'trapped' by the operation of the housing market. Stedman Jones (1971) argued of the late East End of London that there was considerable evidence that the poorest sections of the population were unable or unwilling to move out of the area. Similar arguments have been applied to the role played by the housing market in the employment problems currently experienced by the residents of many inner city areas and peripheral council estates and we shall develop these below.

The second consequence follows from the fact that labour supply tends to be relatively fixed in the short term by residential structure and the operation of the housing market. It means that the local housing market could have a constraining influence upon both employment and labour relocation. As Pahl et al (1983 p.86) have observed: 'For the bulk of potential migrants access to improved job opportunities is inextricably linked with problems over access to housing. Labour market constraints interact with those of the housing market.' When firms establish themselves in an area they rarely 'create' their own workforce out of thin air. For the most part, the potential workforce already exists along with a pre-existing bundle of skills and other attributes. The nature and characteristics of the workforce in any area is a legacy of the developmental history of the local labour and housing markets modified by in and out migration. This iterative historical process of development creates a variety of socially

226

differentiated 'pools of labour' which potential employers are able to tap. This process is broadly similar to that identified by Massey (1979) in her analysis of the role played by different 'rounds' of investment in the shaping of regional economies.

Spatial variations in social stratification can therefore be seen as the outcome of the iterative historical superimposition of changes in the structure of the labour and housing markets in different areas. The changing spatial division of labour clearly has an impact upon local variations in class compositon, but the changing structure of the housing market can also exert an independent and autonomous influence. Local variations in the geographical structure of social stratification are the product of the interaction of both markets in space.

These labour and housing market processes are not just relatively autonomous; they may also operate non-simultaneously in that the periodicity of the dynamic within each market is not necessarily synchronous. Indeed, investment in housing and industrial production has often been broadly counter-cyclical as Parry-Lewis (1965) and Harvey (1978) have argued. In any given area the two markets may thus be either 'in phase' as they were in London and the South East in the 1930s (Bowley 1937; Marshall 1968; Jackson 1974), and as they are today along the M4 corridor, or 'out of phase' as they are in areas of economic collapse. Whilst such phasing differences may not persist in the long term, they may pose considerable short term problems in rapidly growing areas where, for example, the production of new housing may lag behind the growth of employment. Where the infrastructure of housing and workplaces is already in place and where economic decline rather than economic growth is the order of the day the problems of phasing are very different. As Johnson, Salt and Wood (1974 p.15) point out:

> If growth in employment is not evenly distributed between labour and housing markets, those areas which are growing fastest are more likely to experience greater housing demand relative to supply and in-migrants are likely to find more difficulty in obtaining a house. Conversely, areas of economic decline may experience less demand and potential out-migrants, if they own their own homes, may

have difficulty in selling them.

The process of residential restructuring can perhaps be characterised in terms of 'leading' and 'lagging' edges. The leading edges of such restructuring occur where new housing is being developed on a large scale or where processes such as gentrification are reshaping the private housing market of some areas in association with changes in the structure of employment. Although this 'leading edge' restructuring of residential space is essentially an incremental process it can have a rapid and dramatic impact on the housing market opportunities and social structure of specific geographical areas such as the new and expanded towns (Heraud, 1968; Taylor, 1979; Byrne and Parson, 1983) or areas of large-scale new private construction (Craven, 1969; Jackson, 1974). Where employment growth lags behind the rate of new housing construction or where migration is primarily housing related, the short term result may be increased commuting. At the other extreme, on the 'lagging' side of the space economy, areas which have experienced rapid employment decline are characterised by a housing stock and a population which cannot be as easily run down as the local economic base. In these areas the housing market may act as a drag on the ability of the labour force to respond to employment restructuring. Many people are effectively trapped by the housing market in areas of employment collapse.

This is necessarily a crude characterisation of what is, in reality, a complex set of interactions. Both markets are highly structured internally into a hierarchy of specific sub-markets, each of which exhibit their own level of spatial and temporal fixity (Taylor and Thrift 1983). The point however is that changes in either the structure of housing provision, labour market processes or the structure of households will have repercussions on the spatial structure of social stratification.

Housing and Employment Linkages in Cities and Suburbs

The previous discussion was largely abstract in nature. In this section we wish to move from the abstract to the concrete and examine how the two markets interact in practice to produce geographical variations in social stratification. We shall focus on the transformation of the socio-spatial structure of 'city regions' over the last twenty five years, paying particular attention to the growing socio-economic differentiation of the inner cities from the surrounding suburbs and metropolitan rings.

The accelerated de-industrialisation of Britain's inner cities over the past twenty years is now well established (Massey 1979; Fothergill and Gudgin 1982; Danson, Lever and Malcolm, 1980; Danson 1981; Elias and Keogh 1982; Massey and Hagan 1982; Goddard and Champion 1983). The effects of this process are equally well known (Thrift 1979; Friend and Metcalf 1981). The spatial restructuring of industrial production, the growing segmentation of the labour force into primary and secondary sectors and the differential outmigration of the more skilled (Hamnett 1976; Kennett and Randolph, 1978; Pinch and Williams 1983) has left the less skilled, the unemployed and the economically margin-alised 'high and dry on the grey sands of the inner cities' (Cross 1983). One result of this process has been that unemployment rates in both the inner cities and some peripheral council estates have risen much faster than elsewhere (DoE 1983). The segmentation of the labour force between a primary sector which is increasingly associated with non-metropolitan areas and an inner city secondary sector has become more pronounced (O'Connor 1973; Friedman 1983). The partial 'deskilling' of the inner area labour force and the corresponding growth in the numbers and proportions of non-metropolitan skilled workers are aspects of the same process: Labour market segmentation has involved a marked spatial segmentation (Lovatt and Ham, 1984).

Whilst 'redundant spaces' and their associated redundant 'surplus' populations are an ever present feature of capitalist development (Friend and Metcalf 1981; Anderson et al 1983), the last twenty years has witnessed a growing concentration of the long term unemployed, low income peripheral workers and the state supported poor in the inner cities and some peripheral council estates. The gap between these areas and their increasingly economically and socially marginalised populations and the relatively more prosperous metropolitan peripheries and their more highly skilled and affluent workforce is growing wider (Eversley 1976; Pinch and Williams 1983). As Lee (1979) has put it:

> No longer necessary as the spatial mainstay
> of capital, the big city is redundant. Cities
> are, increasingly, the location of surplus
> labour and, at the same time, the focus of
> the twentieth century socialisation of the
> conditions of nineteenth century industrial
> urbanisation. Whole areas of cities are now
> irrelevant to the expanded reproduction of
> capital(p.61).

It would be wrong however to conclude, as
Danson (1981) does for example, that the concentration of the less skilled and the unemployed
in the inner cities is solely a product of structural shifts in the labour market. Whilst it would
be untenable to argue that the housing market can,
in some way, 'create' unemployment or poverty,
the geographical structure and functioning of the
housing market can and does have a major influence
in the form of socio-spatial stratification which
has accompanied the current phase of metropolitan
restructuring.

The Role of the Housing Market in the Production and Maintenance of Metropolitan Socio-Spatial Segregation

The case for emphasising the role played by the
housing market in the geographical concentration
of the less skilled in the inner cities and the
more highly skilled in the suburbs and metropolitan
peripheries is a strong one. Put at its simplest
it involves just three linked propositions. The
first is that access to the different housing
tenures is unequally distributed. Whilst access
to owner occupation is controlled primarily by
price and income mediated by the availability of
mortgage finance, access to the council sector
is controlled by the waiting list, the points system
and the allocational procedures. The result, as
is well known, is that the housing market in Britain
is socially stratified by tenure. The higher the
socio-economic group, the greater the proportion
of owner occupiers and the lower the proportion
of council tenants and vice versa. The extent of
this socio-tenurial polarisation has increased
over the past 25 years as the socially mixed privately rented sector has contracted, as the less
skilled have increasingly gained access to council
housing and as skilled manual groups have increasingly gained access to owner occupation (Hamnett,

1984a).

The second proposition is that the distribution of housing tenures is also highly geographically differentiated between the inner cities, on the one hand, and the suburbs and metropolitan peripheries, on the other. This is largely the result of the timing of their development allied to marked changes in the dominant prevailing tenure form of new housing production. Thus, whereas the inner cities were almost exclusively built for private renting in the nineteenth century, the growth of owner occupation in the interwar and postwar periods was predominantly a suburban phenonemon. Whilst both the interwar and the postwar years have seen the growth of large scale suburban council estates (see Broadbent, Down and Lansley 1984; Broadbent 1985), the post war years have also witnessed equally large-scale inner city council redevelopment schemes. The result has been the growing segregation of housing tenures on an increasingly large geographical scale.

The pattern of tenurial segregation has not, however, been a uniform one. On the contrary, the form of the segregation varies considerably from one part of the country to another and from one city to another depending on their individual building history, social composition and political control. Thus, although council housing is now the largest single tenure in inner London and owner occupation predominates in outer London (Young and Kramer, 1978, Hamnett and Randolph, 1983). it is clear that in many large Scottish cities such as Glasgow, Edinburgh and Dundee the principal concentrations of council housing are found not in the inner city but in a broken ring around the suburban periphery. Redfern's (1983) comparison of the inner and outer wards of the 19 biggest cities in England and Wales outside London using 1981 census data also revealed considerable complexity and local variety.

The third proposition which follows from the first two, is that the postwar period has seen the growth and intensification of socio-spatial tenurial polarisation in some metropolitan areas. After national changes are taken into account the evidence is clear that whilst all socio-economic groups experienced relative decentralisation between 1961 and 1971, the professional and managerial group decentralised most rapidly and the unskilled and the semi-skilled decentralised least rapidly (Pinch and Williams, 1983). It can be argued that there are two interrelated sets of processes at work. First, there are those processes which are tending to result in the concentration of professional and managerial workers in the metropolitan rings and secondly, there are those processes which are tending to result in the growing relative concentration of the less skilled in the metropolitan cores.

Whilst most attention has been devoted to the second set of processes, both are important in shaping the changing pattern of social stratification within the larger metropolitan areas. The processes involve changes in <u>both</u> the distribution of jobs and houses. Not only has employment tended to decentralise during the last twenty years, but the changing tenurial distribution of the housing stock can itself be argued to be influencing the pattern of intra-regional migration and residential change (Bonnar, 1979, Hamnett, 1986). Variations in the tenure and price structure of the housing market play a major role in shaping the distribution of housing opportunities over space. An increasing number of owner occupiers are both living and working in the metropolitan rings and many live in the rings and commute to work in the urban cores. The changing geographical structure of tenures provides both a set of choices and constraints. As Pinch and Williams (1983, p.145) point out:

> Together with shifts in the location of employment, the decentralisation of skilled manual workers is again likely to have been determined by changing patterns of residence. Rising living standards have meant that many of the more prosperous manual workers have been able to gain access to owner-occupied housing in commuter hinterlands or, alternatively, have been forced by constraints within the housing market to purchase new dwellings near the urban periphery.

The resultant changes in the distribution of the workforce may also be exerting an influence on the form of the spatial division of labour. As Massey (1983a, p.25) has observed in relation to employment changes in the M4 corridor. 'The attraction of the area initially was a combination of accessibility to London and nearness to defence establishments... But since then the place has taken off in another way... The stratum of the

labour force most able to choose where to live...
assume that jobs will follow. And they do'.

Core-ring Polarisation in London and the South East

The three propositions outlined above were discussed
in general terms. In this section we wish
to examine more closely the role played by the
housing market in the production and maintenance
of residential social stratification in London
and the South East during the last twenty five
years. It is well known that the long established
concentration of corporate and financial power
and control in London has, along with the concentration of many other functions - both national
and international - resulted in a high proportion
of managerial and professional jobs and a large
professional and managerial labour force (Waugh
1969). During the course of the 1960s and 1970s
this section of the labour force grew rapidly in
both the South East and the rest of the country,
whilst the skilled, semi-skilled and unskilled
manual labour force declined. Within the South
East, however, a considerable restructuring in
the spatial structure of social stratification
took place over this period. The analysis of inter-
censal change data for the period 1961 to 1981
reveals that the professional and managerial sections of the labour force became increasingly concentrated - both relatively and absolutely - in
the South East outside London whilst the smaller
numbers of the less skilled became increasingly
relatively concentrated within inner London
(Hamnett, 1986).

A sharply defined pattern of tenurial segregation has also developed between inner London and
outer London and the rest of the South East during
the course of the last 25 years. In 1961 the proportion of council tenants in inner and outer London
was very similar at 19 and 17 per cent respectively.
By 1981 inner London's proportion had more than
doubled to 43 per cent whilst in outer London it
had increased just 6 percentage points to 23
per cent. Meanwhile, the proportion of owner
occupiers in outer London increased from 53 to
62 per cent. In the rest of the South East outside
London the proportion of owner occupiers increased
similarly from 53 per cent in 1966 to 64 per cent
in 1981. Of a total increase of almost a million
owner occupiers in the South East during the period

233

1966 to 1981 London gained 200,000 or 20 per cent compared to an increase of 750,000 of 48 per cent in the rest of the South East. The South East outside London accounted for 80 per cent of the regional increase in owner occupation. At the same time, the proportion of council tenants in the rest of the South East increased only marginally.

How far and in what ways are these two trends interrelated? Bonnar (1979) suggested on the basis of his analysis of migration data that the differential tenure and price structure of the housing market within the region played a major role in 'orchestrating' the social differentiated nature of migration flows. Analysis of the interrelationships between changes in tenurial and socio-economic structure over the period (Hamnett 1986) tends to confirm this. When the spatial differences in tenure structure are combined with the constraints on access to owner occupation, the result has been that the growth of peripheral owner occupation has been primarily confined to the managerial and professional groups. The growing council sector in London has also accommodated a growing proportion of the less skilled who were previously housed in the privately rented sector. Geographical differences in the tenurial structure of housing opportunities can be shown to have played a key role in the structuring of residential social stratification.

The Role of the Housing Market in the Concentration of the Less Skilled and Unemployed in the Inner Cities

If the decentralisation of more highly skilled jobs and the geographical concentration of new build owner occupation in the metropolitan periphery go some way to explaining the differential out-migration and concentration of the more highly skilled in such areas, how is the concentration of the less skilled and unemployed in the inner cities to be explained. There are a number of radically different explanations for the higher rates of unemployment in the inner cities. Despite their considerable differences in approach and orientation, it is significant that they all stress the distribution of the housing stock and the operation of the housing market as a major factor.

At one extreme, Metcalf and Richardson (1980) argue that the differences in the level of unemployment from one London borough to another are primarily a product of varitions in the skill and other characteristics of the labour force. These variations in turn are seen as a product of the structure of the housing market from one borough to another. They are quite explicit about this: 'the problem of labour market disadvantage in London is caused...by the given housing stocks' (p.202). They argue that the less skilled individuals who suffer labour market disadvantage:

> live disproportionately in the inner city because that is where the largest stock of cheap housing (especially public housing) is found. The problems are a function of the housing stock accumulated over the past 150 years. Further, the problems become circular, because boroughs whose residents tend to suffer labour market disadvantage build a large amount of public housing in an effort to look after the welfare of their residents. This accounts for the temporal stability of unemployment within London (p.202).

Metcalf and Richardson's borough level regression analysis is open to considerable criticism on a number of counts (Bramley, 1979, 1980; Sayer, 1982), not least because it ignores the crucial role of demand deficient unemployment. As the Lambeth Inner Area Study (LIAS) (DoE 1977, p.89) pointed out: 'vulnerability to unemployment...is not the same thing as unemployment; it only becomes so when the demand for labour falls'. The Lambeth study also stressed the role of the housing market in the analysis of inner city unemployment. Where they differed from Richardson and Metcalf was in arguing that the principal causes of inner city unemployment were to be found in both the falling aggregate demand for less skilled labour and in the existence of an occupational and geographical 'mismatch' between labour supply and demand. They argued that there were two aspects to this 'mismatch'. The first was the fact that whilst the number of professional and managerial jobs had increased, the number of less skilled jobs had decreased. The second aspect was the existence of what they termed the 'housing trap' which effectively prevented the less skilled from following the more skilled in escaping from Lambeth and inner London generally in search of job opportunities elsewhere. As the LIAS put it:

> The choices available to them are owner-occupied houses, which they cannot buy, and public housing, to which they are denied access. Few can go to Outer London boroughs because the authorities there have discouraged public housing for Inner Londoners and, lacking the skills in greatest demand, they have largely been excluded from the new and expanding towns. Thus at present the low-skill workers of Inner London are trapped behind an almost impenetrable cordon (DoE 1977, p.44).

Bramley (1979, 1980) has developed the 'entrapment' thesis more generally. He argues that whilst the concentration of unemployment and low incomes in the inner cities stems in part from spatially specific deficiencies in labour demand caused by the local job losses and decentralisation, it is too simplistic to interpret the existence of secondary labour market characteristics solely in terms of the characteristics of local employment. It is rather, a "function of patterns of residential location which derive from the operation of the housing market" (Bramley 1979, p.69). Constraints on mobility arising from the institutional structure of the housing market are particularly acute in the council sector where mobility is restricted not just by the existing distribution of the stock but by the fact that each local authority operates their own waiting list and allocations policy. Gleve and Palmer (1978) have shown that even standardising for social class, council tenants are much less likely to move across their local authority or regional boundaries. The ability of council tenants to commute on a day to day basis to remote or awkwardly located work opportunities is also restricted by their generally low levels of car ownership (Hamnett 1983). This problem can be compounded by the contraction of the local job market or by the decentralisation of jobs (see Kain 1968; Wheeler, 1969; Harvey, 1973). Although the 'entrapment' thesis has come under attack from Cheshire (1979) and Evans (1980) on the grounds that higher rates of unemployment were characteristic of all occupational groups in the inner city and not just the less skilled, they do not challenge the importance of the housing market. Thus Cheshire makes a major distinction between the determinants of the spatial structure of unemployment differentials at the inter- and intra-regional levels. Whereas he sees the former as resulting from the unevenness of industrial structure, 'The relative stability of intra - urban unemployment differences...probably results mainly from the logic of urban structure and patterns of residential location' (1979, p.32). This reinforces the idea that the housing market has a greater degree of locational 'fixity' than the labour market.

The strong linkages which exist between the growth of the urban surplus population and its concentration into what remains of the privately rented sector, marginal inner-city owner occupation (Karn, 1969) and, most importantly, the council sector has been the subject of considerable attention of late. The fear that the council sector is becoming relatively 'residualised' and limited to the less skilled, the economically inactive and the economically marginalised has been a central theme in this work (Taylor 1979; English 1982; Byrne and Parson, 1983; Forrest and Hurie 1983; Malpass 1983). Nor is the problem confined to the inner cities. Byrne and Parson (1983, p.141) argue that the surplus population or, as they term it, the 'surplus reserve army', is also found in 'many new towns and overspill estates peripheralised by de-industrialisation in the very recent past'. Although the fact that, for historic reasons, council housing tends to be concentrated in precisely those areas which have been hardest hit by the current crisis is a contingent rather than a causal relation, the importance of the links between residualisation in both the labour and housing markets cannot be understated. The segmentation of the labour market is becoming increasingly tied to a segmentation of the housing market and the combination of changes in both markets is resulting in a growing degree of socio-tenurial segregation, both between the inner areas and the peripheral rings, and also within them.

Conclusions

Geographical variations in social stratification cannot be understood solely as a product of the changing spatial division of labour, crucial though that is. Nor can they be understood solely as a product of the structure and operation of local labour markets. As we have attempted to demonstrate, spatial and temporal variations in social stratification are a product of the superimposed, iterative and contingent interaction of both labour and housing markets in time and space. Indeed, the concept of local labour markets can only be defined in terms of the existence of a spatially defined pool of labour - and hence in the context of a local housing market. Although each market is autonomously constituted, each market to some extent defines the other. There is a reciprocal, though not necessarily equal, relationship between the two markets, and the nature of this relationship varies both spatially and temporally.

One of the major dimensions of spatial variation in social stratification is the geographical scale of influence of the two markets. Although the structure of employment is arguably the more important determinant of spatial variations in social stratification at the national and regional scale, the role of the housing market has always been important at the local scale. Its importance has increased as improvements in transportation and the growth of commuting have increased the spatial scale of labour market/housing market interaction and reduced the work place constraints on residence for the better off and more mobile.

The degree of spatial and temporal 'fixity' in the two markets is another important dimension of variation. Not only are the two markets not necessarily congruent, but their relative fixity has varied over time. The 'employment linkage' was arguably the most important determinant of residential social stratification during the mass urbanisation of the nineteenth century. Whereas the high cost and limited availability of urban transportation imposed tight constraints on residential location, the high rates of inter- and intra-urban migration were assisted by the dominance of private renting. Today, however, with the growth of suburbanisation and counter-urbanisation (Hamnett and Randolph 1982), the replacement of private renting by council housing and owner occupation and the onset of de-industrialisation, it can be argued that the 'housing linkage' is becoming increasingly crucial. Once the housing stock of an area is in place, a large proportion of the population in declining areas characterised by a high proportion of council housing and or low house prices, tend to be more spatially tied to the area than either industry or employment. This fact is of major importance given the growing tendency towards the segmentation of the labour market and the growing socio-spatial polarisation of the housing market into two socially and spatially

238

distinct tenures. These two tendencies appear to be related. Whereas owner occupation is increasing catering for the more skilled sections of the economically active population, the council sector would seem to be increasingly catering to the economically and socially marginalised.

With the concentration of the unemployed, the less skilled, and other economically marginal groups in some inner city areas, low quality peripheral council estates cannot be interpreted solely in terms of local employment decline. It also results from both the tenurial, price and allocational structure of the housing market, the resultant pattern of social and economic differentiation and the 'frictional drag' which housing exerts on the mobility of labour and the spatial structure of the housing market which divides inner and outer areas in terms of skill and income characteristics. The development of metropolitan housing markets leads not only to the spatial structuring of class and income in residential terms. It also forms a relatively rigid spatial framework which only slowly adapts to changes in the spatial structure of economic activity. The low income secondary and unemployed labour force is concentrated in inner cities not just that is because where they work, or used to work, but because that is also where they live.

References

Anderson, J. Duncan, S. and Hudson, R. (1983) Redundant Spaces in Cities and Regions, Academic Press, London.

Aspinall, P.J. (1982) 'The internal structure of the housebuilding industry in nineteenth century cities' in Johnson, J. and Pooley, C. (eds.) The Structure of the Nineteenth Century British City, Croom Helm, London, pp. 75-105.

Ball, M. (1981) 'The development of capitalism in housing provision', International Journal of Urban and Regional Research, 6, pp. 145-177.

Berry, B. (1970) 'The geography of the United States in the year 2000', Trans. I.B.G. 51, pp. 21-54.

Beshers, J.M. (1962) Urban Social Structure, Free Press of Glencoe.

Bonnar, D. (1979) 'Migration in the South East of England: An Analysis of the Inter-relationship of Housing, Socio-Economic Status and Labour Demand', Regional Studies, 13, pp. 345-359.

Bowley, M. (1937) 'Some regional aspects of the Building Boom', Review of Economic Studies, pp. 172-86.

Bramley, G. (1979) 'The inner city labour market' in C. Jones (ed.) Urban Deprivation and the Inner City (1979) Croom Helm, London, pp. 63-91.

Bramley, G. (1980) 'Unemployment problems in Lambeth' in A. Evans and D. Eversley (eds.), The Inner City: Employment and Industry, Heinemann, London, pp. 271-293.

Broadbent, A. (1985) 'Estates of another realm', New Society, 14 June, pp. 410-411.

Broadbent, T.A., Down, F. and Lansley, S. (1984) 'Outer estates in Britain: Interim report' CES Paper No. 23, CES Ltd., London.

Byrne, D. (1984) 'Dublin: a case study of housing and the residual working class', International Journal of Urban and Regional Research, pp. 402-420.

Byrne, D and Parson, D. (1983) 'The state and the reserve army: the management of class relations in space' in J. Anderson, S. Duncan and R. Hudson (eds.), Redundant Spaces in Cities and Regions, Academic Press, London, pp.127-154.

Castells, M. (1977) The Urban Question, Edward Arnold, London.

Cheshire, P. (1979) 'Inner areas as spatial labour markets: a critique of the inner area studies', Urban Studies, 16, pp. 29-43.

Cooke, P. (1981) 'Tertiarisation and socio-spatial differentiation in Wales', Geoforum, 12, pp. 39-53.

Cooke, P. (1982) 'Class interest, regional re-structuring and state formation in Wales', International Journal of Urban and Regional Research, 6, pp. 187-203.

Craven, E. (1969) 'Private residential expansion in Kent', Urban Studies, 6, pp. 1-16.

Cross, M. (1983) 'Racialised poverty and reservation ideology: blacks and the urban labour market' paper given at the Fourth Urban Change and Conflict Conference, Clacton-on-Sea, 4th-6th Jan.

Danson, M.W. (1981) 'Industrial structure and labour market segmentation': urban and regional implications', Regional Studies, 16, pp. 255-65.

Danson, M.W. Lever, W.F. and Malcolm, J.F. (1980), 'The inner city employment problem in Great Britain, 1952-76: a shift-share approach', Urban Studies, 17, pp. 193-210.

Dennis, N. Henriques, F.M. and Slaughter, (1957) Coal is our Life, Eyre and Spottiswoode, London.

Department of the Environment, (1979a) Policy for the Inner Cities, HMSO, London.

Department of the Environment, (1979b) Inner London: policies for dispersal and balance, Final report of the Lambeth Inner Area Study, HMSO, London.

Duncan, S. (1979) 'Qualitative change in human geography', Geoforum, 10, pp. 1-4.

Dyos, J. (1968) 'The speculative builders and developers of Victorian London', Victorian Studies, 11, pp. 641-690.

Elias, P. and Keogh, G. (1982) 'Industrial decline and unemployment in the inner city areas of Great Britain: a review of the evidence', Urban Studies, 19, pp. 1-16.

English, J. (1982) The Future of Council Housing, Croom Helm, London.

Evans, A. (1980) 'A portrait of the London labour market' in A. Evans and D. Eversley (eds.) The Inner City: Employment and Industry, Heinemann, London, pp. 204-232.

Evans, A.W. and Richardson, R. (1981) 'Urban unemployment: interpretation and additional evidence', Scottish Journal of Political Economy, 28, pp. 107-24.

Eversley, D. (1976) 'Aspects of measuring the social and economic effects of outward movement from metropolitan areas', Paper presented at LARUS Conference, Hamburg.

Forrest, R. and Murie, A. (1983) 'Residualisation and council housing', Journal of Social Policy, 12, pp. 453-468.

Frankenberg, R. (1966) Communities in Britain, Penguin, Harmondsworth.

Freidmann, A. (1983) 'Social relations at work and the generation of inner city decay' in J. Anderson et al (eds.) Redundant spaces in cities and regions, Academic Press.

Friend, A. and Metcalf, A. (1981) Slump City, Pluto Press, London.

Giddens, A. (1973) The Class Structure of Advanced Societies, Hutchinson, London.

Giddens, A. (1979) Central Problems in Social Theory

Macmillan, London.

Gillespie, A.E. and Owen, D.W. (1981) 'Unemployment trends in the current recession', Area, 13 p. 184-96.

Gleve, D. and Palmer, D. (1978) 'Mobility of labour: are council tenants really handicapped', CES Review, 3, pp. 74-77.

Goddard, J.B. (1983) 'Structural change in the British space economy' in J. Goddard and A. Champion (eds.) The Urban and Regional Transformation of Britain, Methuen, London, pp. 1-26.

Haddon, R. (1970) 'A minority in a welfare state society', New Atlantis 2 pp. 80-13.

Hamnett, C. (1976) 'Social change and social segregation in Inner London, 1961-71' Urban Studies, 13, pp. 261-271.

Hamnett, C. (1983a) 'The conditions in England's inner cities on the eve of the 1981 riots' Area, 15, pp. 7-13.

Hamnett, C. (1983b) 'Regional variations in house prices and house price inflation', Area, 15, pp. 97-109.

Hamnett, C. (1984a) 'Housing the two nations: socio-tenurial polarisation in England and Wales, 1961-81', Urban Studies, 43, pp. 389-405.

Hamnett, C. (1984b) 'The postwar restructuring of the British housing and labour markets: a critical comment on Thorns', Environment and Planning, 16 pp. 147-161.

Hamnett, C. (1984c) 'Gentrification and residential location theory: a review and assessment' in Geography and the Urban Environment, 6 D. Herbert, and R. Johnston, (eds.), John Wiley, pp. 283-319.

Hamnett, C. and Randolph, W. (1982) 'The changing population distribution of England and Wales, 1961-81, Clean break or consistent progression' Built Environment, 8, pp. 272-280.

Hamnett, C. and Randolph, W. (1983) 'The changing tenure structure of the Greater London housing market, 1961-81', The London Journal 9, pp. 153-164.

Harris, R. (1983) 'The spatial approach to the urban question: a comment' Environment and Planning, Society and Space, 1, pp. 101-105.

Harvey, D. (1973) Social Justice and the City, Edward Arnold, London.

Harvey, D. (1978) 'The urban process under capitalism: a framework for analysis', International Journal of Urban and Regional Research,

2, pp. 101-131.

Heraud, B.J. (1968) 'Social class and the new towns', Urban Studies, 5, pp. 33-58.

Jackson, A. (1974) Semi-detached London, Allen and Unwin, London.

Johnson, J.H. Salt, J. and Wood, P.A. (1974) Housing and the Migration of Labour in England and Wales, Saxon House, Farnborough.

Kain, J. (1968) 'Housing segregation, negro unemployment and metropolitan decentralisation', Quarterly Journal of Economics, 83, pp. 175-93.

Karn, V. (1979) 'Low income owner occupation in the inner city' in C. Jones (ed.), Urban Deprivation and the Inner City Croom Helm, London, pp. 160-190.

Kemp, P. (1981) 'Housing and landlordism in late nineteenth century Britain', Environment and Planning, 14 pp. 1437-47.

Kennett, S. and Randolph W. (1978) 'The differential migration of socio-economic groups, 1966-71', Discussion Paper No. 66, LSE Graduate School of Geography.

Kirby, A. (1983) 'On society without space: a critique of Saunder's nonspatial urban sociology', Environment and Planning Society and Space, pp. 226-233.

Lee, R. (1979) 'The economic basis of problems in the city' in D.T. Herbert and D.M. Smith (eds.), Social Problems and the City: Geographical perspectives, Oxford University Press, pp. 47-62.

Lipietz, A. (1980) 'Inter-regional polarisation and the tertiarisation of society', Papers of the Regional Science Association, 44, pp. 3-17.

Lovatt, D. and Ham, B. (1984) 'Class formation, wage formation and community protest in a metropolitan control centre', International Journal of Urban and Regional Research 8, pp. 354-387.

Malpass,P. (1983) 'Residualisation and the restructuring of housing tenure', Housing Review, March/April pp. 1-2.

Marshall, J.L. (1968) 'The pattern of housebuilding in the interwar period in England and Wales', Scottish Journal of Political Economy, 15, pp. 184-205.

Martin, R. (1982) 'Britain's slump the regional anatomy of job loss', Area, 14, pp. 257-64.

Massey, D.B. (1979) 'In what sense a regional

problem?, Regional Studies, 13, pp. 233-43.

Massey, D.B. (1983a) 'The shape of things to come', Marxism Today, April, pp. 18-27.

Massey, D.B. (1983b) 'Industrial restructuring as class restructuring: production decentralisation and local uniqueness', Regional Studies, 17, pp. 73-90.

Massey, D.B. (1984) Spatial Divisions of Labour: Social Structures and the Geography of Production, Macmillan, London.

Massey, D. and Meegan, R. (1982) 'Industrial restructuring versus the cities', Urban Studies, 15, pp. 273-80.

McGregor, A. (1979) 'Area externalities and urban unemployment' in C. Jones, (ed.), Urban Deprivation and the Inner City, Croom Helm, London, pp. 92-112.

Metcalf, D. and Richardson, R. (1980) 'Unemployment in London' in A. Evans, and D. Eversely (eds.) The Inner City: Employment and Industry, Heinemann, London, pp. 193-203.

Noyelle, T.J. (1983) 'The rise of advance services', American Planning Association Journal, (Summer) pp. 280-290.

O'Connor J. (1973) The Fiscal Crisis of the State, St. Martin's Press, New York.

Pahl, R. (1965) 'Trends in social geography' in R.J. Chorley and P. Haggett (eds.) Frontiers in Geographical Teaching, Methuen, London, pp. 81-100.

Pahl, R. (1970a) 'Spatial structure and social structure', in R. Pahl, Whose City?, Longmans, London, pp. 185-198.

Pahl, R. (1970b) 'Urban social theory and research' in Whose City?, Longman, London, pp. 209-226.

Pahl, R. (1970c) Whose City?, Longman, London.

Pahl, R. Flyn, R. and Buck, N.H. (1983) Structures and Processes of Urban Life, Longman, London.

Palmer, D. and Gleve, D. (1981) 'Employment, housing and mobility in London', London Journal, 7 pp. 177-193.

Parry-Lewis, J. (1965) Building Cycles and Britain's Growth, Macmillan, London.

Payne, J. and Payne, G. (1977) 'Housing pathways and stratification: a study of life chances in the housing market', Journal of Social Policy, 6, pp. 129-56.

Pinch, S. and Williams, A. (1983) Social class change in British cities' in J. Goddard, and A. Champion, (eds.) The Urban and Regional

Transformation of Britain, Methuen, London. pp. 135-159.

Redfern, P. (1983) 'Profile of our cities', Population Trends, 30, pp. 21-32.

Rex, J. and Moore, R. (1967) Race, Community and Conflict, Oxford University Press, Oxford.

Santos, M. (1977) 'Society and space: social formations as theory and method', Antipode, 9, pp. 49-59.

Sassen-Koob, S. (1984) 'The new labour demand in global cities', in M.P. Smith (ed.) Cities in Transformation, Vol. 26, Urban Affairs Annual Reviews, Sage, Beverly Hills, pp. 139-171.

Saunders, P. (1980) Urban Politics: A Sociological Interpretation, Penguin, Harmondsworth.

Saunders, P. (1981) Social Theory and the Urban Question, Hutchinson, London.

Saunders, P. (1984) 'Beyond housing classes: the sociological significance of private property rights and means of consumption', International Journal of Urban and Regional Research, 8, pp. 202-227.

Sayer, A. (1979), 'Theory and empirical research in urban and regional political economy: a sympathetic critique', Urban and Regional Studies Working Papers, 14, University of Sussex.

Sayer, A. (1982) 'Explanation in economic geography: abstraction versus generalisation', Progress in Human Geography, 2, pp. 68-88.

Simmie, J. (1982) 'Beyond the industrial city?' Journal of the American Planning Association, 49, pp. 59-76.

Stedman Jones, C. (1971) Outcast London, Oxford University Press.

Taylor, P. (1979) 'Difficult to let, difficult to live in and sometimes difficult to get out of', Environment and Planning, A, 11, pp. 1305-20.

Taylor, M. and Thrift, N.J. (1983) 'Business organisation, segmentation and location', Regional Studies, 17, pp. 445-465.

Thrift, N.J. (1979) 'Unemployment in the inner city: urban problem or structural imperative?' in D. Herbert and R. Johnston, (eds.) Geography and the Urban Environment, vol. 2, John Wiley, Chichester, pp. 125-226.

Urry, J. (1981) 'Localities, regions and social class', International Journal of Urban and Regional Research, 5, pp. 455-73.

Vance, J. (1966) 'Housing the worker: the employment linkage as a force in urban structure', Economic Geography, 42, pp. 294-325.

Waugh, M. (1969) 'The changing distribution of professional and managerial manpower in England and Wales between 1961 and 1966', Regional Studies, 3, pp. 157-169.

Wheeler, J. (1969) 'Transportation problems in the negro ghettoes', Social Science Research, 53, pp. 71-9.

Young, K. and Kramer, J. (1978) Strategy and Conflict in Metropolitan Housing, Heinemann, London.

CHAPTER TEN

SOCIAL RELATIONS, RESIDENTIAL SEGREGATION AND THE HOME

PETER WILLIAMS

Whatever the country it would seem most people are acutely aware of the social divisions within their society. The terms may vary (as between class, income, owners/non-owners) but the sense of social groupings and social divides comes through clearly. Nowhere is this more apparent than in discussions of residential areas. The residents of any town or city have a well formed awareness of the status of any area reflecting the types of housing and their occupants. Again the no-menclature may vary but posh, mixed, up and coming, downmarket, exclusive and desirable are just some of the many adjectives used to denote social status and social trajectory. Amateur sociologists and social geographers abound, yet whilst we all may have an intuitive grasp of the complex strati-fication of space, detailed understanding of the way this has come about, and more generally the way social relations in total are played out in space is remarkably limited. Academic research may have pointed to the patterns which result and their regularity between different cities and cul-tures, but our comprehension of how these patterns evolve and change is weak, and our understanding of the way social relations are composed and con-textualised in space (Thrift 1983) is at best rudimentary.

Nowhere is this more true than when we turn to 'the home'. Increasingly within social science, reference is made to the spread of privatisation, home centredness, consumption classes and the importance of the rise of home-ownership; yet a close consideration of the literature reveals only limited attempts to open out and consider the home around which so much discussion now focuses.

This chapter has two main aims. The first

is to argue the need to recognise weaknesses in
existing analyses which, while giving more attention
to non-work based relations, are still short of
any substantial focus on the home. The second is
to sketch out ways in which we might conceptualise
the home as a site and situation where social
relations, including class, status and gender are
composed and contextualised. The chapter aims
to achieve these aims by briefly reviewing current
literature on stratification, segregation and the
housing market which, in essence, is the work most
closely related to this theme. After this the
chapter then moves on to review work on the home
and to layout a broad framework for further development. As will be evident this is an exploratory
paper reflecting work in progress (see Saunders
and Williams 1984; Williams forthcoming).

Stratification, segregation and the housing market

Within mainstream sociology and geography considerable attention has been given to the division of
residential space into stratified territories and,
to a lesser extent, to the impact that these have
had on local social relations. By extension, housing and homes are implicated in the process, but
though the former has been developed substantially
the latter remains neglected. In part, this reflects
the powerful and continuing impact of the Chicago
School. The view of cities as a series of rings
or sectors and the confusing theatrical terminology
of the dress circle and the stalls continues
strongly in almost every urban sociology or social
geography text book. An acceptance of the pattern
(even when it does not parallel reality) has also
led to an often unstated acceptance of how it is
assumed to come about - a preference for space
which when related to income ensures a gradient
of social classes and social locations (poor in
centre, rich on the periphery).
      Concern with stratification in space - resi-
dential segregation - has been dominated by an
interest in form and appearance rather than cause
and consequence. The social division of labour
between geography and sociology has not assisted
overcoming this problem. Within sociology there
has been a tendency to present residential segregation as a simple manifestation of stratified
societies. Thus inequality in work, and class
manifest by work, are stamped upon the social landscape in terms of where people live and how they

live. Certainly other arguments have been developed and these have now gained momentum (witness collective consumption sectors, housing classes, gender and identity), but the intensity of the debate on class and the concern with the workplace has meant that residence and residential segregation have been low priorities. The focus on the workplace rather than the residence and the neglect of the interlinkages between them has until recently relegated reproduction/consumption issues to a secondary role. Thus, where it is mentioned in books on class and inequality, residence and segregation are typically seen as a reinforcement process (see for example, Abercrombie and Urry 1983, p.24). Where consumption has been treated more seriously (e.g. Dunleavy 1980; and Saunders 1984), it is clear that there is evidence of a crosscutting of work-based class relations by consumption (especially home ownership) but that there is uncertainty as to how and why this is occurring. Significantly a number of commentators are pointing to the home and home-ownership as crucial issues in these processes of apparent dealignment (see Saunders 1978, 1984). We will return to this later.

Geography has been obsessed with spatial patterns and processes (see Massey, forthcoming, for a review) and has sought explanation for spatial variation within a very limited explanatory range. Location is distilled into a trade-off between space and cost with social conflict being reduced to a simple bidding process. Thus, despite the Chicago School's concern with ethnography, social interaction and the social and moral order (Jackson 1984), geographers typically rendered this invisible and emphasised only the inevitable concentric zones. Even now, despite the emergence of stronger theorised geographies it is common to find residential segregation being accepted as given rather than explained. These are general tendencies. In the past few years there have been encouraging developments within both disciplines in terms of the capacities to engage in and develop a much better understanding of the social world. With regard to residential segregation, class and spatial relationships there is wider recognition of the unsatisfactory nature of our present approaches and the need to begin to come to terms with the ways in which class and other relations may actually be constituted outside the workplace (see Urry 1981; Cannadine 1982; Harris 1984; Jackson and Smith 1984; Saunders 1984.

This process has been substantially assisted by a cross disciplinary interest in housing and the housing market which has offered substantial insights to both the housing market and its relationships to stratification and segregation. Thus, work on access and opportunity, constraints and urban managers, housing and collective consumption, the state and ideology have all contributed to an understanding of the way housing is an active social process which reflects, reinforces and redirects processes within society related to both production and consumption. In other words, our understanding of a person's housing has now shifted from being a simple container of family life to one which raises questions about allocative devices, commodity production and distribution, finance capital and exchange professionals, ideology, political practice and incorporation (see Short 1982 for a review of British material and Ball 1983 for a more developed account of the political economy of housing).

The broadening and deepening study of housing and housing markets has been revealing and important. It has, however, tended to develop a momentum which removes it from consideration in relation to other social arenas. In other words, it has tended to become a separate debate with its own dynamics. Moreover, because of the particular intellectual origins of many of the central arguments (marxist economics), debates around housing have (until recently) been quite remote from issues of the home as a site and setting for the constitution of social relations. In general terms, many of the analyses are dominated by a somewhat mechanistic and fragmented sense of the nature of housing and its place in the social world. Typically, one is led through accounts of different tenures, their costs and subsidy implications with the market being presented as an allocative device sorting households across the urban landscape. The heritage of the ecologists is clear to see in that there is still a considerable focus on the spatial outcomes of the housing system. More recently this restrictive distributional understanding of housing has been extended by a concern to establish the basis by which housing is produced and consumed - thus a concern with housing as a commodity, commodity production and a contextualised understanding of market institutions and professionals.

This apparent coming together of the pro-

duction, distribution and consumption of housing
has led some to assume that our understanding of
housing is now 'complete'. That however is far
from the case. The formulation is dominated by
a sense of housing as an economic relation and
indeed even the communication of politics under
this approach is almost reduced to a sense of a
formal cost-benefit assessment of specific tenures.
In short, while we have extended our grasp of housing as an economic relation we have gone hardly
anywhere in terms of its role as a social relation.
Certainly there has been an active debate on ideology - the tenant as potential revolutionary, the
owner as determined conservative, but it has always
been tinged by an economistic perception of the
matter in hand (see the concern with capital gains,
the process of commodification and the issues around
the stake in the system).

The reaction to the 'limited' political economies of housing, class, politics and gender has
begun to gather considerable strength giving both
the space and support for extending and developing
arguments in different directions. The question
now is should the home be part of that development? In the section that follows we briefly consider some of the existing arguments related to
the home and then we seek to extend these into
a more structured whole as the basis for future
work.

The Home

Despite the actual or implied importance ascribed
to the home our comprehension of this 'locale'
is extremely limited and in some respects quite
distorted. The most common view of the home arises
with respect to questions of privacy, status and
identity reflecting both a received wisdom drawn
from daily life and academic research (see Mellor
1977; Perin 1977, Duncan 1981). None of these
issues has been closely explored yet all are frequently referred to. Beyond these concerns we
are left with a largely uncharted territory crossed
by some known paths (e.g. women and the built environment and feminist research in general,
sociology of the environment, domestic technology,
social history and urban ethnography), most of
which have developed in isolation of each other.
Clearly there are interconnections and taking the
home as a focus provides one basis upon which to
explore this complex mass of material. Seeking

to maintain a concern with processes and relations there would appear to be a number of separate but related themes through which we can begin to comprehend the home as an active social relation. These themes are privacy and control, status and identity, politics and political change, gender relations and social relations and commodification. It must be made clear it is not claimed these are the only processes with which we should be concerned. However, a reading of diverse literatures and careful reflection does suggest that these encapsulate and begin to integrate some of the arguments already before us which, although not always linked to the home, clearly have an important bearing upon our understanding of the 'locale'.

Privacy and Control

Richards in his book The Anatomy of Suburbia (1973, p.13) has commented:

> Ewbank'd inside and Atco'd out, the English suburban residence and the garden which is an integral part of it stand trim and lovingly cared for in the mild sunshine. Everything in its place. The abruptness, the barbarities of the world are far away.

As the quote indicates, the suburban home is presented as a private sanctuary offering its occupants the control and stability denied at work and in residential locations closer to the city centre. The suburbs are supposedly remote from class antagonism, pollution and chaos. Their role is to regenerate the occupants by providing space, calm, control, convenience and congeniality. The facts of production, as Williams (1983) put it, were to be banished and insecurity replaced by the deeply rooted security of the home fire. Thus alongside privacy, there has been a natural focus on the family, upon security and stability. It has been assumed that, in contrast to inner areas, suburbs are unchanging, offering long term viability and the basis for 'solid' family development. Suburban children not only grew up in a safe environment but also a homogenous one, learning the values and mores of a single social group, or so the imagery runs. Living in a new environment meant the possibility of good facilities (if and when built) and less need or likelihood of the close interwoven neighbour networks which may have existed elsewhere.

All of these descriptions of suburbia have some truth to them though it must be said that research evidence is often contradictory and there is a high degree of cultural specificity. Perceptions of privacy and security are not the same for all people nor is the extent of family/home centredness. Moreover, the model of suburbia used is very middle- income and middle- class/homeowning. In Britain, there are many working-class renting suburbs and for some occupants (not just renters) economic insecurity is not left behind at the front door. Indebtedness, unemployment, disrepair all ensure that the home for some is no longer a haven - it can be a trap. The failure to meet housing bills, whether for rent or mortgages, can result in eviction and repossession (a phenomenom increasingly common in the UK in the 1980s). The pressure to become the owners of a dwelling, and often a dwelling which stretches the resources of a household is such that the slightest downturn in income can result in a rapid collapse of the flimsy household economy and ownership. The imagery conjured up by Richards (1973, p.35) in the quote below is thus not of the generality he may have wished it to be:

> The first instinct, however, of the suburban dweller is his craving for economic security and in a world that does not provide this, for a defence that is always close at hand. That he is able to take refuge from the implications of such knowledge in the domestic charms of a manufactured environment is a tribute to his typically English capacity.

While the typical model is a suburban home it is apparent that commentators regard any home (especially home-owned homes) as a locus of privacy and control. In part, this imagery is sustained by a contrast with work. Work and the workplace is public - penetrated by numerous influences and forces - home and the residence is private-controlled, exclusive and protected. The view of home as private is sustained through design, convention, class and personal practice. The physical representation of the home in British society whether as a flat or a semi is of a unit enclosed by hedges, walls, net curtains and doors removing it from the public domain, however close it may be. The privet hedge is a perfect symbol of this process. Convention too makes much of this separation. People apologise for ringing you at home as if somehow this sanctuary is not to be penetrated. People often make a point of never going inside a neighbour's home because it implies an intimacy and familiarity which is somehow unacceptable. Such practices have strong class connotations.

Although control may be expressed through privacy this is not an inviolable relationship. The form of control households wish to exercise relates strongly to factors such as age and class. Young people often lead a highly communal existence reflecting the desire to separate themselves from parental control (tempered by economic realities). Here privacy is less important than control itself. Middle-aged and especially middle-class households take a different view of the situation. There is, it seems a clear tendency to demand more space, and privacy as they grow older. Affluence enables some to achieve this and in Britain it may be expressed in detached and semi-detached houses. Later on in life as family needs change and isolation grows, elderly people often seek to return to a more communal existence with a carefully controlled link between privacy and public oversight.

Privacy provides households with opportunities to control and exclude. But the home is more than that. Within the home, households have opportunities to express themselves as individuals and through that to emphasize their class position and ambitions. The capacity to individualise is now sustained through the provision of a wide range of goods and services which seem to expand on an hourly basis. As people gain control of their dwelling space (through the reduction in household size and to some extent the growth in regulated ownership and renting) so the opportunities to express are both sustained and created. Being able to do what one wants when one wants and how one wants would be seen as a critical factor with regard to perceptions of housing as a home.

This emphasis on privacy, exclusion and control can easily be overemphasised. The home, the dwelling place, is not totally isolated. It is linked via legal, physical and economic instruments to the outside world. Homes are serviced by external agencies, subjected to external control and penetrated by numerous external forces. The telephone, television, gas, electricity, drainage, planning, noise, pollution are just some of the many ways in which this so-called private sphere is rendered public. Certainly in relative terms the household may have more control there than in the shop, factory or library but there is no absolute control. The home as castle is substantially an ideological device.

Related to the matter of control is the issue raised by Giddens (1981) and pursued by Saunders (1982, 1984) of 'ontological security'. Giddens uses this term to describe the security, continuity and meaning which surrounded life in tribal and class divided societies but which is now being undermined in capitalist society. He argues that commodification and the deconstruction of time and space have removed meaning from people's lives. The growth of home-ownership, Saunders argues, may be a particular expression of the desire to re-establish ontological security. Until such time as we have substantially more empirical evidence it will be hard to challenge these arguments. It is clear from existing work on the meaning of the home, on housing and identity and housing and social status, however, that there are considerable complexities. Rakoff (1977, p.85) comments:

> houses are used to demarcate space, to express feelings, ways of thinking, and social processes and to provide arenas for culturally defined activity as well as to provide shelter.

In interviews Rakoff conducted with households it was apparent that houses were seen as commodities, places for child rearing, indicators of status and success, bases for permanence and security, and refuges and that to achieve any of these ownership was necessary. Rakoff argues that the variety of meanings reflects the ambiguity of the private sphere, individualism and the ways in which each individual household is responding to and recreating social and economic structures. Thus in Vancouver, Pratt (1981) shows the very distinct forms of meaning two different groups of women attached to their homes and sought to express through it - the aspirant middle class seeking individuality while the established wealthy group sought to conform to the canons of good taste. These studies of the home emphasise the importance of ownership (see also Deverson and Lindsay 1975; Holdsworth 1979). However, just as meanings vary amongst owners so other meanings arise amongst tenants, some not dissimilar to owners. For the

255

tenant, as with any occupant, the home still offers a degree of control (see for example Hoggart 1958, Pahl 1984). The home, in these terms, may mean the possibility of growing one's own vegetables, decorating a room, cooking or simply being together with someone else. Certainly the added bonuses of capital gains and a high degree of freedom of action give owners advantages just as the landlords rights give tenants disadvantages but the advert which stated (cited in Holdsworth 1979, p.192), 'The house you live in is not home if you don't own it. If you are paying rent you are living in someone else's home, not your own home is wrong. That home is someone else's commodity not their home'. Regardless of accuracy this is a powerful message and would be repeated by many.

Status and Identity

It is already apparent that the home, as both a physical structure and location and as a socio-psychological concept, is a factor in terms of the status and identity of its occupants, reflecting and reinforcing other social processes and con-ditions such as class and income. Burnett has asserted that the home (1978) 'is the most signifi-cant mark of social differentiation and the most significant symbol of social status' though it is not immediately clear on what basis he makes the claim. Intuitively we may be tempted to agree although, in Britain for example, accent, dress and family background would appear equally signifi-cant. Furthermore, since many households do not occupy the dwelling of their choice it would be unwise to ascribe an automatic relationship between occupant and home. Indeed as Duncan (1981), Steinfeld (1981) and others make clear the meaning attached to housing in terms of status and identity varies significantly across societies and within societies. Housing can be enabling and disabling and through that process communicate very different images of its occupants. Steinfeld's research on the elderly demonstrates this well, revealing that for many social groups, societal definitions as to their position substantially curtail the housing opportunities available to them. The posi-tion of young single people is not dissimilar. Equally other groups who might be categorised as "the norm", such as the so called typical nuclear family might also have quite restricted housing opportunities, e.g. in Britain, the three bed semi-detached house.

Whatever the significance of the home with respect to status and identity, it would appear to derive out of a number of characteristics namely the dwelling type and design, its builder and architect, its location, previous occupants and history, the contents and amenities of the dwelling and the social signification given to all of these attributes in the society concerned. The title of Oliver, Bentley and Davis' interesting book on the semi-detached house provides an appropriate starting point. Dunroamin: the Suburban Semi and its Enemies is a valuable account of suburbia and the semi-detached house tracing through the pressures for and against the creation of suburbia to the dwelling type which came to characterise it - the semi. The popular image of the semi (and suburbia) as dull uniformity contrast markedly with the intentions underlying its creation. The semi was part of the vision of individuality which expressed itself in suburbia (Richards, 1973). At one level the semi is remarkably uniform but, as Bentley (in Oliver et al 1981) demonstrates, the range of choices confronting the householder were staggering in contrast to their previous dwellings...Dwelling size and structure varied to a small extent but major features such as windows, bays, gables and porches were offered in extensive permutations as were stained glass windows, fireplace styles, panelling, kitchens and bathrooms. Add to this the 'planes of choice' related to domestic equipment and furnishings and the massively symbolic detail of garden furnishings, pictures etc., and the semi ceased to be anything other than an intimate expression of the occupants, their life style and aspirations. There is not space in this chapter to pursue this matter in greater detail but the point is made - the popularity and dislike of the semi reflecting only too clearly the imagery of status and identity it carried. To the aspirant working and middle-class it was highly desirable. To the established middle-class and the upper-class it was effrontery and had to be denigrated.

Dwelling type and design, as Mumford (1961) makes perfectly clear, carries with it numerous messages regarding the occupants. It signifies position and direction, and by doing so, acts to demarcate the occupants from those who occupy different types of dwellings. The hierarchy of dwelling types, from terrace, end terrace, semi-detached to detached, marks in England a careful social trajectory (allowing for variations such as the Georgian terrace or town house)(1). Similarly there are flats and mansion flats as even within a single type there is hierarchy of which estate agents are the self-appointed guardians. Building types and more particularly designs move in and out of fashion. Since this is reflected in market price there almost always comes a point where the out of fashion design/type returns to fashion climbing through the ranks as good value to highly desirable before it reaches overpricing. None of this is static - Georgian terraces were unfashionable in the early post-war period - suburban semis are now becoming fashionable. As society's values are reshaped and changed reflecting economic, social and political pressures, so this is expressed in a variety of ways including dwellings. But it is more than a 'passive' reflection of what exists or is becoming. The pressures on the housing market may restrict the choices open to particular groups such that having made them they must then be justified. The homes industry is well capable of dealing with that through the myriad of magazines, property columns and agents.

Just as building type and design can ascribe status to an occupant so can its builder or architect. It is astonishing to find houses in certain streets valued more highly than others because a particular builder constructed the dwellings. Somehow this is then used to indicate that the occupant is more discerning (typically this issue is made more important where the house design is commonplace e.g. the semi, and a new criteria for demarcation becomes important and relevant). A similar process is apparent with respect to historic properties and areas - the present occupants accruing status from their decision to live somewhere with historical associations. In a more dramatic form gentrification has a similar meaning. Restoring a Victorian workingman's cottage to its former glory is taken to mean discernment (i.e. not following the herd into new houses), thrift (such houses often better value in terms of space), self sufficiency (typically the owner engages in DIY) and a respect for the past (including a rejection of mass production, a love of natural finishes and an appreciation of good workmanship). In combination gentrification and the gentrified house is a visible expression, nay celebration, of a whole set of middle-class values (see Jager, 1986 for an interesting paper on this process).

The promotion of status through the home and the associated commodification of this process so that societal values are translated into monetary equivalents is, of course, plainly apparent in the jostling for the right address. Location is all important in British society. In a complexly structured society the mechanisms for indicating the subtle yet crucial distinctions between individuals and groups are all important. One's address, which becomes involved in numerous daily transactions, communicates an image of the occupant of that address which relates to status, wealth and ambition. London postal districts are a perfect illustration of this. Of course living at the right address requires the capacity to bid for that location. In a competitive housing market, a price differential is paid for the better addresses as people bid and jostle for the right locations. Definitions of the right address may derive from earlier historic associations, now cemented in the form of the buildings extant there, proximity to other facilities and amenities (parks, schools, rivers, views etc.) or even important neighbours. While so called fashionable addresses may rise and fall many consolidate and are quite stable. Moreover although there are many areas which never achieve any notable status, within each local housing market there is normally a very precise hierarchy of addresses/areas (Deverson and Lindsay 1975). Of course this process is given even more obvious stature and meaning by the acutely aware naming of streets and houses. Titles such as The Grove, The Rise, The Old High Street often indicate streets of distinction; add to that names such as 'The Old Hall', 'The Schoolhouse', 'The Bakery' and one begins to form mental images of dwelling and occupant. Over time the range of titles has been extended to Avenues, Drives and Closes which summon up images of development in the thirties and the fifties, while Glenroy, Bertilda, Dunroamin perfectly illustrate the aspirants desire to mark out a territory and a castle using the style of the upper classes (see Betjeman on Metroland, in Delaney 1983). The social patterning of British towns can, in part, be read off a map in terms of street names and it is perhaps surprising how often First, Second and Third Avenue really do mark out a social gradient with different building and occupant types.

The home, in the terms described here, is

without doubt a significant indicator of status. How far it is a determinant is difficult to determine. The relationship is certainly more than one way since the imagery conjured up by the right address, dwelling or whatever, influences and effects a myriad of other relationships. A 'bad' address can mean no credit, a requirement for extra phone installation payments, difficulties in obtaining services and all manner of other problems. Equally the right home can lubricate relationships and insert the occupants in social networks which can have profound effects in individual and class terms. The home can thus consolidate, strengthen and it may even initiate. We return to this issue later.

Politics and Political Change

The variety of meanings of the home have already been discussed, yet within contemporary social science there has been an intense and rather polarised debate regarding the political significance of the home (see Saunders 1984 for a review, Kemeny 1980). The home-owned home has been claimed to generate a certain protective conservatism on the part of its occupants which may, in turn express itself in a vote for Conservative governments. The converse has been suggested for the occupants of council-rented dwellings where the collective nature of the arrangement, in conjunction with the dependent relationship established with respect to the local authority and central government, is seen to encourage votes for Labour governments. Certainly the evidence of home-owners voting Conservative and council tenants voting Labour exists but that in itself is not sufficient evidence to claim the relationship is an inviolable reality. Questions of ontological security and personal control, of capital gains and unearned income certainly endow the home-owned home with significant meaning for many households (Forrest, 1983). Depending upon other circumstances, the home may stand as a central pillar of a household's present and future prospects and as such be something which is a key concern when voting in elections and in terms of their support for particular structures of social and economic provision.
    The debate on home-ownership relates to the broader and longer running debate on embourgeoisement and dealignment within the British class structure. The decline of the close association

between working -class voters and the Labour party
has caused many analysts to seek out the bases
for any new cleavages which might be emerging.
Dunleavy (1981), Saunders (1984) and others have
placed considerable stress on the declining significance of the workplace and class (as defined by
position at work) as the driving forces within
British society. Instead, they argue, consumption
has become far more important and given rise to
sectoral cleavages which cross cut and even displace
class in terms of providing the bases upon which
people act. Saunders argues that one such sectoral
cleavage is between home-owners and non-home-owners.
As he states (1984, p.207).

> housing tenure, as one expression of the divi-
> sion between privatised and collectivized
> means of consumption, is analytically distinct
> from the question of class...such cleavages
> are in principle no less important than class
> divisions in understanding contemporary social
> stratification, and because housing plays
> such a key role in affecting life changes,
> in expressing social identity and...in modify-
> ing patterns of resource distribution...it
> follows that the question of home ownership
> must remain as central to the analysis of
> social divisions and political conflicts.

According to these arguments, the home, both rented
and owned homes, though especially the latter,
are seen to have critical impacts on political
behaviour in its broadest sense. Certainly part
may translate into voting behaviour though all
analysts would now agree the relationship is not
as simple as once argued. For many ownership is
not so much a device for capital accumulation and
incorporation within the Conservative party but
rather a device through which independence and
freedom of action can be achieved and the capacities
to stand outside of the normal wage relationship
(Rose 1981; MacKenzie and Rose 1983). Ownership
or non-ownership can influence the options open
to a household and the choices made. The opportunity to accumulate wealth through ownership may
result in a household having the capital to buy
health, education and even legal services. The
denial of these opportunities, as must be the case
with renters, can entrap a household and diminish
life chances.
 There is some evidence to support Saunders

arguments. However, it must be recognised that there are many owners whose political behaviour and ideology is less influenced by their consumption position than their situation in the labour market. This is certainly the case with many state employed professionals who, despite their possession of domestic property, have swung substantially behind the Labour party in recent years. This, of course, links to arguments about the proletarianisation of the middle class which in turn indicates something about a shift in the meaning of home-ownership. Defence of turf (Cox 1984) becomes less important than defence of the job which pays for it. The substantial rise in defaulted mortgages, indeed the growth in the (unreported) numbers of householders who simply abandon ownership and hand in the keys, reflects the fact that, in the scheme of things, households place jobs above homes. Exceptions can easily be found. The resistance to leaving the mining village or steel town after the closure of the plant reflects for some households an absolute determination to hang onto what is their own and what is familiar. In reality, both groups exist in large numbers and the precise way any single household will go will depend upon their own circumstances.

It becomes apparent that the home can be an important factor in politics and political change. Through its capacity to influence identity, opportunity and networks the home can, in a variety of ways and situations, have an influence on political action. It has been argued that ownership is crucial to this process and certainly that it may enhance particular tendencies. But surely of equal importance is the question of security and control. A household secure in its home (in legal and economic terms) has in essence a different horizon over which to view the political scene. Denied that security, as private rented tenants are (and increasingly public sector tenants), then political choices are narrowed, though in all cases this must be seen in relation to other equally crucial determinants like jobs and income. The home not only influences politics - it is part of politics and struggle over housing and the home has been a continuing feature of British political life.

Gender Relations and Social Relations

The question of gender relations and social

relations with respect to the home is perhaps one of the most complex and contentious areas for discussion. The far-reaching and vitally important debates within feminist literature with regard to productive and unproductive labour, women and design, gender and class, and the home are crucial issues which are not easily resolved here(2). Feminists have rightly taken issue with assumptions in the literature regarding women in the home. Typically, as the term 'housewife' demonstrates, the home is seen as a private sphere remote from work and supervised by a women. This so called service role has been grossly misunderstood not least because even now 'housework' is treated as non-work and women as inferior to men. As the discussion has already indicated the home is no simple and passive container, and this is true with regard to gender relations and, beyond that, social relations in general and class relations in particular.

Feminist debate has exposed the weaknesses of many common assumptions re the home. More to the point, it has indicated the importance of the home as one locale where gender and class relations are forged. For men, the home may well be a setting for command and consumption while for women it is often obedience and production (Raphael, 1983). Obviously this is not universally so and all women exercise control in a variety of ways. However, a male as wage earner, as the person locked into what are often the more rapidly changing and challenging forces operating in the workplace may often treat the home as 'his domain' (his castle?) where his choices and needs are paramount. For a women working at home a degree of dependence is induced by the denial of a wage (housekeeping money is not a wage) and the requirement to work at maintaining the home and the household within it. Even in households where a more equalitarian relationship exists it is striking how the responsibilities for the home (at least in terms of cleaning, food purchasing and preparation, washing and tidying) have been sustained. Men may undertake some or all of these task but women typically have prime responsibility. Since work can define status and class it is unsurprising that many women, unable to gain recognition for their own (house) work, may be forced to seek an identity via the wage earner. This situation reinforces their dependence upon the income earner and diminishes control over their own lives.

It would be easy to develop an account which merely reinforced this view of women as trapped into an inferior situation - it clearly exists. However, while one can find numerous examples of this process - note even the division of labour re the interior and exterior of the home (interior - women's work; exterior men's work) the situation is far more complex. It is striking that women play a very significant role in decisions to buy/rent a particular property, its interior decoration and furnishing and in the maintenance of social relations via the home. Male obsessions with the workplace often diminish their capacity to think of the home except along one dimension - how much will it cost. The female is more used to having to deal with crucial locational questions such as access to schools, shops, neighbours and services (see, for example, Women and Geography Study Group, 1984), matters of design (Matrix, 1984) and decoration, furnishing and equipping. While some men may be skilled in matters of refurbishment it would seem women are the central actors in many residential location decisions.

Selecting homes is important but women also take a crucial part in managing and operating the home as a social unit. Because of 'duties' such as taking the children to school women are often very quick to establish contact with local social networks. Having done so they also assume control of them, filtering children's friendships and adult social engagements. In doing so women are crucially influencing processes of status and class constitution. Given their role in selecting locations in which they live (albeit within some cost constraint over which they may have less control) and this management function, it could be argued, women (and the gender division of labour) have a significant impact on class relations (if, as it must be, these are seen as being constituted through the totality of production and reproduction). Denied access to the regime of paid work, many women assume a highly significant role in non-work social relations. The implication of this is that, if the role of the work in the constitution of class is declining and other spheres are becoming more important, then women may be more obviously central to these constitutive processes.

Of course the home is no isolated castle. The home is a physical location and as an interaction structure, is penetrated by a myriad of social forces. The design of the dwelling itself reflects and reinforces class, status and gender divisions. The kitchen, for example, is displaced to the rear of the dwelling reflecting not only a class division in the era of servants but also a view regarding the place of domestic work and women. The corridor, the servants stairs, trades-man's entrance were just some of the devices used to allow persons of different class and status to move around a dwelling without interfering with the principal occupants (see Williams, forthcoming; Mumford 1961). Oliver goes further with his essay on the home as a human symbol. He comments (1981, p.161).

> The feminine forms of typical semis, before which the front garden was spread like a trim apron, had connotations of the mother as home-maker so important in the period. The woman's place was not only in the home; the home was a woman.

Such concerns illustrate some of the ways in which the world beyond shapes and changes relations within. Television, radio and the media in general are, of course, not repulsed by the front gate, nor can one ignore the daily forays made by women, children and men into the arenas of school, work, shopping and leisure, all of which bring back to this setting a variety of influences and pressure.

The point then is that relations within the home are dynamic and subject to constant change. While external forces are clearly important, members of a household are active participants in the process as 'intermediaries' and as key actors. Within the home the relationships between men and women and parents and children as always being negotiated and fought over. Attitudes, images and actions are always being subjected to scrutiny, reinforce-ment and change. While this may have been seen by some as 'mere' consolidation of external forces such as class there is a debatable line between consolidation and composition. If lives are becom-ing more privatised and households more home centred, if the world of formal work is being changed (and transposed via technology into the home), then the home itself and all that goes on there must be becoming more important as a locale where social relations are forged and changed.

Commodification, Accumulation and the Home

Although we have sought to open out the home and to consider the processes operating there it is essential not to lose sight of some of the broad market relations which surround the home and intimately influence its form and function. Mention has already been made of the debate regarding home-ownership and capital accumulation. In this section we briefly review the links between the home, commodity relations and capital accumulation.

The connection between the home-owned home and money is made very strongly in all advertisements and inducements. Decisions about what home to buy and what to spend on it when bought are influenced by views regarding monetary gains. In other words at one level we are acutely aware of the connections. Less apparent but equally important are the ways in which the whole process by which homes become commodities reaches out into society as a whole (see Forrest and Williams, (1984) for a full discussion of this). In Britain, homes are produced for profit. Housing is an expensive and essential element for all persons though the opportunities to obtain decent housing are quite restricted. Around this nexus of housing as a need and housing for profit have grown a whole panopoly of institutions, policy instruments and private enterprises all competing and often contradicting one another. Housing costs influence wages and the division of profits on society so everyone would claim an interest in how housing questions are resolved.

The profits derived from housing come in a number of different forms. Physical construction involves, land, materials, labour and finance and each has become an important market. Once constructed, the home requires equipping with furnishings and appliances and again that is a major market. Fashion, obsolescence and simply use, results in a high turnover of domestic goods and major efforts are made to keep pressure on households to re-equip. The same is true of refurbishment, extension and adaption. The home is a personal space and as such can be adapted and changed to meet individual needs however minor. All of these interests mean that what you and I do in our homes (and to our homes) is a matter of great importance and numerous enterprises are always seeking to shape and influence our action. Homes are sites of consumption as well as being places produced by the building industry and the failure to provide or create space for, say washing machines, does influence prospects for other industries.

In one way or another homes are sources of profit. In themselves they are also stores of wealth. The home-owned home allows the householder to make realisable capital gains which can be translated into the purchase of goods and services. Promoting home-ownership is seen as highly beneficial even though, in the short term, purchase is normally crippling to a household's expenditure patterns. All of these different implications mean that a home, separate from its occupants, has great meaning in market societies. It is itself a commodity, capable of releasing monetary gains, and it influences consumption patterns in a whole variety of ways.

Finally, it is important not to overlook the home as a site of production and thus as a place from which profitable enterprises may operate. Homework is increasing substantially reflecting unemployment, technological change and the growth of small businesses. Paid work can be undertaken in the home ranging from an executive and her workstation, through the self-employed plumber or electrician to the home knitter or Christmas cracker maker. The design of the home and the capacity to control the home environment in an appropriate way are obviously necessary prerequisites for any significant production base (so too are planning regulations and other external controls).

Concluding Comments

One aim of this chapter has been to awaken our understanding of the home as a setting, a locale where a variety of forces interact. It is apparent that the home, that secluded private sphere, is anything but a passive container where tranquility and peace prevails. In a myriad of different ways the home stands central to contemporary social relations and, if one follows certain arguments, its importance is increasing. How then can social scientists persist in simply making gestures towards an understanding of this sphere as they have done for so long. As we learn more about class, residential segregation or social status it is apparent we must begin to look more closely at the home as one of the settings in which, potentially at least, important interactions take place.

It is increasingly clear that simple arguments regarding the paramount importance of variable X or Y are doomed to failure. Social relations are complex and variable, and though it is held that methodologically and conceptually we now have the equipment to undertake the task of unravelling relations and constructing adequate theory, we are without doubt hampered by our primitive understanding of so many arenas of social life and the home is one of these.

It has not been possible within the confines of this chapter to make any substantial theoretical advance. That stage is rapidly approaching but as a first priority it seemed essential to sketch out, in a structured way, the relations and processes operating within and around the home. In doing so one begins to question the rather common but slight references made to the home in debates about the decline of work and the role of the work place in class societies. In many such papers the home is presented as the passive realm remote from work and an alternative or counter to it. The home thus becomes a mechanism by which the decline of 'class' is amplified. However, if a somewhat less differentiated view of the world is adopted it could be argued that the home is acting to create and consolidate class relations regardless of the situation in the work place. Moreover in doing so it is also a critical setting with respect to gender and class and class and status. The discussion in this chapter would suggest that the simple separation of the social world into spheres of work and non-work are not helpful since relations and responses are carried back and forth between one and the other through economic necessity, ideology, politics and all the forces that give rise to social action (Thrift, 1983).

Careful consideration of the home makes it difficult to demarcate from work whether in the formal or informal economies. The home situation (form and functioning) is crucially conditioned by the work situation and vice versa. The concern to ensure the workforce is appropriately housed and the desire to exploit that opportunity has resulted in employers having substantial involvement whether it be via company mortgage schemes and subsidy arrangements, the direct provision of housing or pressure on government re-housing provision. Furthermore, as part of the process of managing production relations, employers have structured that assistance selectively promoting particular strata or employees (managers) and, in doing so, have assisted the consolidation of that class and the promotion of social needs (O'Connor, 1984).

Just as position in the labour market may enhance one's housing situation so the reverse is also true in both the short and long term. The location of the home provides a social, political and educational context which, although it may be transcended by existing linkages or the expenditure of income, provides an important setting for the social and economic development of a household and its occupants. In the short-term networks may mean new contracts and jobs, in the longer term these and educational opportunities may result in a rapid enhancement of life chances and social trajectories. The home thus carries with it transformative potential in terms of social relations and social positions.

The significance we attach to the home rather depends on the arguments being developed. A dominant concern with production left little space for an understanding of reproduction and consumption with the consequences that, to some extent, attention to this arena became part of opposing arguments. Recent work within marxist political economy (e.g. Harvey 1982, O'Connor 1984, Preteceille and Terrail 1985) begins to acknowledge the crucial intersections between production and consumption, not least around the question of social needs, reproduction costs, and the value of labour power; and this carries with it, the prospect of different conceptions of the bases of class formation and action. The preliminary arguments of this chapter would suggest the home provides an important setting for both the constitution of class and, ironically, its fragmentation via the numerous mechanisms for differentiation and distancing that arise.

NOTES

1. See Muthesius (1982) for an account of the terraced house and more generally the evolution of the home.
2. The literature is extensive see Hayden (1981), Hunt (1980), Matrix (1984), Women and Geography Study Group (1984). See also the important and interesting literature on domestic technology, (e.g. Vanek 1978; Schwartz-Cowan 1979; Thrall 1982; Bose 1984).

REFERENCES

Abercrombie, N. and Urry, J. (1983) Capital, Labour and the Middle Class, Allen and Unwin, London.
Ball, M. (1983) Housing Policy and Economic Power, Methuen, London.
Bose, C. et al, (1984) 'Household technology and the social construction of housework', Technology and Culture, 25, pp. 53-82.
Burnett, J. (1978) A Social History of Housing, David and Charles, Newton Abbot.
Cannadine, D. (1982) 'Residential differentiation in nineteenth century towns: from shapes on ground to shapes in society' in J. Johnson and C. Pooley (eds.), The Structure of 19th Century Cities, Croom Helm, London pp. 235-252.
Cox, K. (1984) 'Social Change, turf politics, and concepts of turf politics' in A. Kirby, P. Knox and S. Pinch (eds.) Public Service Provision and Urban Development Croom Helm, London, pp. 283-315.
Delaney, F. (1983) Betjeman Country, Paladin, London.
Deverson, J. and Lindsay, K. (1975) Voices from the Middle Class Hutchinson, London.
Duncan, J. (1981) 'From container of women to status symbol: the impact of social structure on the meaning of the house' in J. Duncan (ed.), Housing and Identity, Croom Helm, London, pp. 36-59.
Dunleavy, P. (1980) Urban Political Analysis, Macmillan, London.
Forrest, R. (1983) 'The meaning of homeownership', Society and Space, 1, pp. 205-216.
Forrest, R. and Williams, P. (1984) 'Commodification and housing: emerging issues and contradictions', Environment and Planning 16 1163, 1180.
Giddens, A. (1981) A Contemporary Critique of Historical Materialism, Volume 1, Power Property and the State, Macmillan, London.
Harris, R. (1984) 'Residential segregation and class formation in the capitalist city', Progress in Human Geography, 8, pp. 26-49.
Harvey, D. (1982) The Limits to Capital, Blackwell, Oxford.
Hayden, D. (1981) The Grand Domestic Revolution, MIT Press, Cambridge, Mass.
Holdsworth, D. (1979) 'House and Home in Vancouver: images of West Coast Urbanism, 1866-1929' in G. Stelter and A. Artibise (eds.), The

Canadian City, McClelland and Stewart, Toronto, pp. 186-211.

Hoggart, R. (1958) The Uses of Literacy, Penguin, Harmondsworth.

Hunt, P. (1980) Gender and Class Consciousness, Macmillan, London.

Jackson, P. (1984) 'Social disorganisation and moral order in the city' Transactions, Institute of British Geographers, 9, pp. 168-180.

Jager, M. (1986) 'Class definition and the aesthetics of gentrification: Victoriana in Melbourne' in N. Smith and P. Williams (eds.), Gentrification of the City, Allen and Unwin, London.

Kemeny, J. (1980) 'Home ownership and privatization', International Journal of Urban and Regional Research, 4, pp. 372-388.

MacKenzie, S. and Rose, D. (1983) 'Industrial change: the domestic economy and home life' in J. Anderson et al (eds.), Redundant Spaces, Academic Press, London, pp. 155-200.

Massey, D. (1985) 'New directions in space' in D. Gregory and J. Urry, (eds.), Social Relations and Spatial Relations, Hutchinson, London.

Matrix, (1984) Making Space, Pluto, London.

Mellor, R. (1977) Urban Sociology in an Urbanized Society, Routledge and Kegan Paul, London.

Mumford, L. (1961) The City of History, Secker and Warburg, London.

Muthesius, S. (1982) The English Terraced House, Yale University Press, New Haven.

O'Connor, J. (1984) Accumulation Crisis, Blackwell, Oxford.

Oliver, P., Bentley, I. and Davis, I. (eds.), (1981) Dunroamin: The Suburban Semi and its Enemies, Barrie and Jenkins, London.

Pahl, R. (1984) Divisions of Labour, Blackwell, Oxford.

Perin, C. (1977) Everything in Its Place, Princeton University Press, Princeton.

Pratt, G. (1981) 'The house as an expression of social worlds' in J. Duncan (ed.), Housing and Identity, Croom Helm, London, pp. 135-180.

Preteceille E. and Terrail, J. (eds.) (1985) Capitalism, Consumption and Needs, Blackwell, Oxford.

Rakoff, R. (1977) 'Ideology in everyday life: the meaning of the house' Politics and Society, 7, pp. 85-104.

Richards, J. (1973) The Castles on the Ground: The Anatomy of Suburbia, 2nd edition, Murray, London.

Rose, D. (1981) 'Homeownership and industrial change: the struggle for a separate sphere' Urban and Regional Studies Working Paper No.25, University of Sussex.

Samuel, R. (1983) 'The middle classes between the Wars' New Socialist, Jan/Feb, 33.

Saunders, P. (1978) 'Domestic property and social class', International Journal of Urban and Regional Research, 2, pp. 233-251.

Saunders, P. (1984) 'Beyond housing classes', International Journal of Urban and Regional Research, 8, pp. 202-227.

Saunders, P. and Williams, P. (1984) 'Charity begins at home: some thoughts on recent and future developments in Urban Studies', paper given to the British Sociological Association Sociology and Environment Group, November (to be revised for publication).

Schwartz-Cowan, R. (1979) 'The industrial revolution in the home: household technology and social change in the 20th Century Technology and Culture, 17, pp. 1-23.

Short, J. (1982) Housing in Britain: The Post-War Experience, Methuen, London.

Steinfield, E. (1981) 'The place of old age: the meaning of housing for old people', in J. Duncan (ed.), Housing and Identity, Croom Helm, London, pp. 188-246.

Thrall, C. (1982) 'The conservative use of modern household technology', Technology and Culture, 23, pp. 175-194.

Thrift, N.J. (1983) 'On the determination of social action in space and time', Environment and Planning D, Society and Space, 1, pp. 23-58.

Urry, J. (1981) The Anatomy of Capitalist Societies, Macmillan, London.

Vanek, E. (1978) 'Household technology and social status: rising living standards and status and residential differences in housework, Technology and Culture, 19, pp. 361-374.

Williams, P. (forthcoming) 'A social history of the home' in N.J. Thrift and P. Williams (eds.), The Making of Urban Society, Routledge and Kegan Paul, London.

Williams, R. (1983) Towards 2000, Hogarth Press, London.

Women and Geography Study Group, (1984) Geography

SOCIAL RELATIONS, RES. SEGREGATION AND THE HOME

<u>and Gender</u>, Hutchinson, London.

DAVID BYRNE is Lecturer in Social Policy at the University of Durham. He was previously Reader in Social Studies at the Ulster Polytechnic and Research Director of North Tyneside Community Development Project. Recent publications include a 'A rejection of Gorz's Farewell to the working class', Capital and Class, 24, 1984. He is a Labour Councillor in Gateshead.

JAMIE GOUGH worked in the Industry and Employment Branch of the Greater London Council, and is carrying our research into the development of London manufacturing in the late 1970s. He has worked as a Research Fellow and Lecturer in town planning at Middlesex Polytechnic. He is co-author with Mike Macnair of Gay Liberation in the Eighties.

CHRIS HAMNETT and BILL RANDOLPH are, respectively, Lecturer and Research Fellow at the Social Science Faculty of the Open University. Both authors are currently involved in an ESRC-funded project on the relationship between recent changes in the structure of the housing market and labour markets of the London region. Previous research has included various aspects of housing market and social change, particularly in London and the South East.

KEITH HOGGART is a Lecturer in Geography at Kings' College, University of London. Educated at the Universities of Salford, Toronto and London. His present research interests are local government finance, urban political parties and rural development. He is currently visiting Associate Professor in Urban Studies, Temple University, Philadelphia.

ELEONORE KOFMAN is Senior Lecturer in the School of Geography and Planning at Middlesex Polytechnic. She has published articles on regional development and social movements in France and contributed to Geography and Gender. Her current research interests are recent developments in French social geography and decentralisation in France. From 1983 to 1985 she was the Secretary of the Social Geography Study Group and is an active member of the Women and Geography Study Group.

LINDA PEAKE is Lecturer at the School of Development Studies at the University of East Anglia. She has published articles on feminism and political geography. Her current research is on gender and urban planning in Britain and the Carribean. She is at present Secretary of the Women and Geography Study Group of the Institute of British Geographers.

MONIQUE PINÇON-CHARLOT is a researcher in the Centre Nationale de la Recherche Scientifique at the Centre de Sociologie Urbaine, Paris. She has published numerous articles and books on social space and public services in the Paris region and is currently working on high-ranking civil servants, their social practices and provision of public services in France.

NIGEL THRIFT is currently co-Director of the Centre for the Study of Britain and the World Economy at Saint David's University College, Lampeter. He has carried out research at the University of Cambridge, University of Leeds and the Australian National University. He is the author of numerous books and articles as well as being the co-editor of Environment and Planning A and a member of the editorial board of Society and Space.

JOHN URRY is Professor and Head of Department of Sociology, University of Lancaster. Educated at Cambridge 1966-1970 and Lecturer at Lancaster since 1970. He is the author/joint-author of Reference Groups and the Theory of Revolution, Social Theory of Science; The Anatomy of Capitalist Societies; Capital, Labour and the Middle Classes; Localities, Class and Gender. He is joint-editor of Power in Britain and Social Relations and Spatial Structures.

ALAN WARDE is Lecturer in Sociology at Lancaster University. He has written Consensus and Beyond: the development of Labour Party Strategy since 1945; Contemporary British Society with N. Abercrombie et al; and Localities, Gender and Class with other members of the Lancaster Regionalism Group.

PETER WILLIAMS is Assistant Director, Institute of Housing, London. He previously worked at the Australian National University and the University of Birmingham. He has undertaken research on housing markets in Britain and Australia and has published papers and monographs on housing, social theory, urban change and gentrification. He has edited/co-edited Conflict and Development, Social Process and the City and Urban Political Economy and Social Theory. He is co-author of Public Housing and Market Rents in South Australia and Salvation and Despair: Home Ownership in the Inner City. He is on the editorial boards of Environment and Planning A and Society and Space.

Printed and bound by CPI Group (UK) Ltd, Croydon, CR0 4YY

22/10/2024

01777621-0013